HOT TOPICS

ALTERNATIVE ENERGY SOURCES

THE END OF FOSSIL FUELS?

By Sophie Washburne

Portions of this book originally appeared in *Energy Alternatives* by Karen D. Povey.

LUCENT
PRESS

Published in 2019 by
Lucent Press, an Imprint of Greenhaven Publishing, LLC
353 3rd Avenue
Suite 255
New York, NY 10010

Designer: Deanna Paternostro
Editor: Jennifer Lombardo

Library of Congress Cataloging-in-Publication Data

Names: Washburne, Sophie, author.
Title: Alternative energy sources : the end of fossil fuels? / Sophie Washburne.
Description: New York : Lucent Press, [2019] | Series: Hot topics | Includes bibliographical references and index.
Identifiers: LCCN 2018019073 (print) | LCCN 2018022416 (ebook) | ISBN 9781534565104 (eBook) | ISBN 9781534565098 (library bound book) | ISBN 9781534565081 (pbk. book)
Subjects: LCSH: Renewable energy sources–Juvenile literature.
Classification: LCC TJ808.2 (ebook) | LCC TJ808.2 .W37 2019 (print) | DDC 333.79/4–dc23
LC record available at https://lccn.loc.gov/2018019073

Printed in the United States of America

CPSIA compliance information: Batch #BW19KL: For further information contact Greenhaven Publishing LLC, New York, New York at 1-844-317-7404.

Please visit our website, www.greenhavenpublishing.com. For a free color catalog of all our high-quality books, call toll free 1-844-317-7404 or fax 1-844-317-7405.

CONTENTS

FOREWORD

Adolescence is a time when many people begin to take notice of the world around them. News channels, blogs, and talk radio shows are constantly promoting one view or another; very few are unbiased. Young people also hear conflicting information from parents, friends, teachers, and acquaintances. Often, they will hear only one side of an issue or be given flawed information. People who are trying to support a particular viewpoint may cite inaccurate facts and statistics on their blogs, and news programs present many conflicting views of important issues in our society. In a world where it seems everyone has a platform to share their thoughts, it can be difficult to find unbiased, accurate information about important issues.

It is not only facts that are important. In blog posts, in comments on online videos, and on talk shows, people will share opinions that are not necessarily true or false, but can still have a strong impact. For example, many young people struggle with their body image. Seeing or hearing negative comments about particular body types online can have a huge effect on the way someone views themselves and may lead to depression and anxiety. Although it is important not to keep information hidden from young people under the guise of protecting them, it is equally important to offer encouragement on issues that affect their mental health.

The titles in the Hot Topics series provide readers with different viewpoints on important issues in today's society. Many of these issues, such as domestic abuse and free speech rights online, are of immediate concern to young people. This series aims to give readers factual context on these crucial topics in a way that lets them form their own opinions. The facts presented throughout also serve to empower readers to help themselves or support people they know who are struggling with many of the challenges adolescents face today. Although negative viewpoints are not ignored or downplayed, this series allows young people to see that the challenges they face are not insurmountable. Abusive relationships can be ended, the internet can be navigated safely, and the environment can be saved.

Quotes encompassing all viewpoints are presented and cited so readers can trace them back to their original source, verifying for themselves whether the information comes from a reputable place. Additional books and websites are listed, giving readers a starting point from which to continue their own research. Chapter questions encourage discussion, allowing young people to hear and understand their classmates' points of view as they further solidify their own. Full-color photographs and enlightening charts provide a deeper understanding of the topics at hand. All of these features augment the informative text, helping young people understand the world they live in and formulate their own opinions concerning the best way they can improve it.

The Problem with Fossil Fuels

As of 2018, the world gets 81 percent of its energy from fossil fuels such as oil, coal, and natural gas to heat homes, keep cars running, provide electricity, and manufacture products people use every day, such as plastic and clothing. However, this reliance on fossil fuels creates several problems.

One of the biggest problems with fossil fuels is that the earth has a limited supply of them. As fossil fuels become harder to find, the prices will increase, and eventually, many people will not be able to afford to put gas in their cars or heat their homes. Wars have already broken out over these disappearing resources. Eventually, these resources will run out altogether, and if people do not find alternatives before that point, even the richest people will be left with no way to maintain the lifestyle they have grown used to.

Another large problem with fossil fuels is that when they are used to create energy, they are bad for the environment because they release carbon dioxide, which gets trapped in the atmosphere. This has many negative effects, both on the planet and on its inhabitants. The air becomes harder to breathe, so people and animals get sick more often. The temperature of the planet rises, which creates climate change and sets off many other changes: melting ice caps, the deaths of animals that cannot adapt to their changing environment, unexpected weather patterns that create dangerous conditions such as hurricanes and floods, and more.

A third problem is the cost needed to maintain the fossil fuel industry. Digging into the earth for coal, gas, and oil costs a lot of money, and transporting the fuel all over the world also contributes to the high cost consumers must pay. In contrast,

Burning fossil fuels contributes to global warming and climate change.

renewable energy sources such as wind and water can be found nearly everywhere, so the cost is limited to the buildings or machines that must be built to convert these resources into energy. They do not need to be transported over long distances or dug out of the ground, and they do not damage the environment if they cannot be contained, the way oil spills do.

What Are Fossil Fuels?

Coal, oil, and gas are all found in the earth and were created by it. Coal is formed over millions of years by woody plant parts, such as tree trunks, leaves, and twigs. When a lot of pressure and heat is put on these plant parts over a long period of time, they slowly transform into coal.

Oil is also formed with heat and pressure over millions of years. It is created from water plants and bacteria that died, fell to the bottom of the ocean, and were covered with a layer of mud. As time passed, the mud turned to rock, which put pressure and heat on the dead organisms. In the absence of oxygen, this heat and pressure turned the organisms into what is called "crude oil." Crude oil is taken out of the earth and refined before being used as energy.

When oil is exposed to too much heat, it breaks down into natural gas. Natural gas is also released while coal is in the process of forming. Since the planet creates all of these materials naturally, it is technically possible for the earth to make more of them. However, since the process takes millions of years, the people who are alive today are using them up too quickly for them to be replaced.

High heat and pressure turns plants into coal over the span of millions of years.

Looking for Alternatives

As Americans become aware of the need to find alternatives to fossil fuels to provide energy for all the activities of daily life, there are fears that this awareness may be too late. Even with a new call to action, many industry experts believe there may not be time to ramp up technology quickly enough to support energy alternatives by the time fossil fuels begin to run dry.

Despite fears that alternatives have been slow to catch on, many new energy technologies are gaining momentum, slowly eating away at fossil fuels' energy dominance. These renewable energy industries, such as wind, solar, and geothermal power, are growing quickly, and their costs are dropping, making them easier and cheaper to adopt. This rapid growth is also creating many new jobs; according to *Fortune* magazine, a study published in 2017 by the Environmental Defense Fund's (EDF) Climate Corps program reported that the "solar and wind industries are each creating jobs at a rate 12 times faster than that of the rest of the U.S. economy."[1]

The quest to develop affordable, effective alternatives to fossil fuels may be one of the most challenging technological tasks in human history. Fortunately, scientists all over the world are up to the challenge, working to uncover the next alternative energy breakthroughs. Supporting their efforts are individuals everywhere who do their part to conserve energy and limit their own use of nonrenewable fossil fuels.

Why Is Alternative Energy Necessary?

The first humans needed an energy source for limited reasons: They generally only needed something to keep them warm and cook their food. Wood-burning fires met both of these needs. However, as society advanced, people began needing energy for a variety of additional things, such as creating pottery and metal tools. Fire was still useful, but wood became more in demand as people began using it to build houses, carts, furniture, and other items as well as to fuel their fires. In search of other energy sources, people eventually turned to fossil fuels.

What Is the Energy Economy?

As forests were depleted and people were forced to venture farther seeking fuel, the first systems for finding and distributing energy evolved. This new "energy economy" nourished an industry that bridged the gap between the need for energy and the fuel supply. A trade in firewood developed in which wood was collected, dried, stored, and sold or traded. Firewood had become a resource that had value to both the dealer and the buyer.

At the same time, the effects of a limited supply of wood could be felt. As wood became scarce, prices rose and the threat of running out became real. This situation had a significant impact on developing societies, which author Paul Roberts explained in his book *The End of Oil*:

> *In short, energy had become a strategic resource—a factor in the rise, and fall, of economies and civilizations. Catastrophic fuel shortages were probably quite frequent. Indeed, the march of*

human progress may well have been marked by a series of energy crises that either killed off a particular civilization or helped push it to the next level of technological and economic development.[2]

Energy shortages were especially severe in Europe. By medieval times, rapidly growing populations and new manufacturing industries were consuming huge amounts of firewood. By the 15th century, firewood was so expensive that only the wealthiest people could afford to buy it. As wood supplies ran out, people turned to using coal for fuel. Coal had been used in China as far back as 2000 BC and had been known about in Europe for centuries as well, but Europeans considered it a poor alternative to wood because it produced thick, smelly smoke that burned eyes, blackened clothing, and ruined the taste of food. It was not until wood became scarce that they began using coal for the majority of their energy needs; in spite of its drawbacks, it was plentiful and cheap, and it provided much more energy than wood when burned.

AN INEFFICIENT COAL-POWERED INVENTION

"A typical stove in 1899, according to a study in Boston, burned some three hundred pounds of coal in a week, produced twenty-seven pounds of ash, and required three hours and eleven minutes of attention."

–Bill Bryson, author

Bill Bryson, *At Home: A Short History of Private Life*. New York, NY: Broadway Books, 2003, p. 125

The invention of the steam engine in the 1700s, however, finally showed people how much coal could do. Engines changed the way energy was used. Instead of just being burned to produce energy for cooking or heating, coal could now be burned to produce mechanical energy as well. This mechanical energy could do the same work that previously required the efforts

When the steam engine was invented in the 1700s, demand for coal increased dramatically. These engines powered many things, including trains.

of people, horses, or oxen. With so much available coal, the Industrial Revolution was fueled. Factories sprang up using engines to power the manufacture of many products. Trains and steamships carried people and cargo throughout Europe and the United States. By 1900, nearly 1 billion tons (907 billion kg) of coal was being used each year.

Turning to Oil

Although coal had replaced wood as a readily available and inexpensive source of energy, it still had its drawbacks. Coal was bulky to store and did not burn hot enough for some industrial uses. Additionally, the soot-related problems of illness and pollution remained. Author Bill Bryson described what it was like to live in London, England, at this time:

> *By 1842, Britain was using two-thirds of all the coal produced in the Western world. In London the result was a near-impenetrable gloom through much of the year … So hard was it to find one's way that people not infrequently walked into walls or tumbled into unseen voids …*
>
> *Coal [smoke] was hard on practically everything—on clothes, paintings, plants, furniture, books, buildings, and respiratory systems. During weeks of really bad fog, the number of recorded deaths in London could easily increase by a thousand. Even pets and animals at the Smithfield meat market died in disproportionately increased numbers.*[3]

Coal's dominance as an energy source changed forever when the first high-producing oil well was drilled in Texas in 1901. Almost immediately, oil became a favored energy source over coal due to its ability to burn cleaner, its ease of storage, and its ability to provide greater amounts of energy. However, because of its abundance and low cost, coal is still widely used, especially in developing countries.

To take oil from the ground, holes called oil wells are drilled, and pumps similar to the one shown here remove it from the earth.

Today, energy—especially that produced by oil—dominates the world's modern cultures. As Paul Roberts observed, "We produce and consume energy not simply to heat and feed ourselves, to move ourselves, or to defend ourselves, but to educate and entertain ourselves, to expand our knowledge, change our destiny, construct and reconstruct our world, and fill it with stuff."[4] Just as the uses of energy have expanded, so has the demand for vehicle fuel and electricity. Global energy use has risen steadily over the past several decades and continues to rise. The World Energy Council has predicted that world energy demand per capita, or per person, will peak in 2030, which means it will reach its highest level and begin to fall after that. However, "while overall per capita energy demand would begin to fall, demand for electricity would double by 2060, the council said."[5] Because of this, countries will have to put more effort into developing energy-efficient technology.

Planning for the Future

Will there be enough oil to meet this growing demand? Many experts say that just as shortages of wood resources caused previous societies to turn to coal, oil shortages will force people to find other ways to meet the world's energy needs. However, exactly when this will happen is up for debate. The oil company BP stated in 2014 that there was enough oil left to last another 53 years, but in 1981, it was predicted that there was only enough oil left to last 32 years, so it is clear that these estimates are not reliable. Financial experts have pointed out that the way oil reserves are calculated is flawed, leading to these confusing statistics. An investment website called the Motley Fool explained,

> *Every country has a slightly different way of calculating the amount, but the basic gist is that proved reserve estimates are what companies assume they can pull from the ground using existing technology while still generating a profit, which is based on the price of oil or gas over the past year.*
>
> *This estimate can be impacted by several things: new technology that makes oil recovery less expensive, new reservoirs that were not considered economical at the time, or simply that a rising price of oil makes those once unattainable resources attainable.*[6]

In other words, when people talk about how much oil is left, they mean how much it will be profitable for a company to remove from the ground. New technology has allowed oil supplies once considered impossible to reach to have increased potential for being tapped. Oil can now be drawn out of deposits of sticky tar sands in Canada, from below frozen tundra, or from deep undersea wells. Tom Ahlbrandt, former project chief of the World Energy Project of the United States Geological Survey (USGS), has an optimistic opinion about these hidden oil supplies. He said, "We feel that more than half of all undiscovered resources are in the deep off-shore, of which half are in the Arctic … We haven't even begun to discover all the oil that is out there."[7] As prices for oil rise, it becomes profitable for companies to spend more money to tap difficult oil sources. At some point, the oil that can be refined and used for energy will be used up, and although the earth can eventually make more, it will take thousands or millions of years. However, as the Motley Fool pointed out, "It is much more likely … that we will reach the economic limits of extracting oil from the ground before we reach the physical limits."[8] Kenneth Rogoff, a Harvard University economist, said, "We might be running low on $20 [per barrel] oil, but for $60 we have adequate oil supplies for decades to come."[9]

If Rogoff is correct, the question then becomes, how much longer will the average person be able to afford to put gas in their car and heat their house with oil-powered energy? Additionally, will society be prepared for a world without fossil fuels? According to Roberts,

> *Yet thus far, governments, and the populations that elect them, seem to be in a state of denial about petroleum. It is true that efforts have been made to develop alternative fuels or shift the energy economy to natural gas, but such programs will cost trillions of dollars and require decades to carry out. Thus, the real question is not whether oil is going to run out (it will) but whether we have the capacity, the political will, to see that outcome soon enough to prepare ourselves for it.*[10]

Preparing an energy-dependent world for a future without oil is a top concern for many people. Many experts fear, however, that progress toward alternative energy sources has been far too slow. Some world governments, including the United States, have put far more money into looking for more oil reserves than into developing alternative energy technologies. Some individuals and private companies have taken the matter into their own hands, but their resources are limited without government support.

No matter what happens, one thing is certain: At some point, people will no longer have access to oil to supply them with energy. This is why many researchers are focusing on alternative energy sources that can be used when oil runs out—although many say the sooner society can decrease its reliance on fossil fuels, the better.

Damage to the Environment

The threat of running out of oil is only one factor inspiring the pursuit of energy alternatives. As the environmental impacts of using fossil fuels become ever more apparent, maintaining the long-term health of the planet has become a driving force for advocates

Coal mines such as the one shown here have negative impacts on the environment as well as the people who work in them.

of finding cleaner energy alternatives.

Some of the environmental costs of extracting, transporting, and burning fossil fuels are visible and have been studied for decades. Reaching fossil fuels buried deep underground often requires major disturbances to natural landscapes. Most coal, for example, is removed through a process known as surface mining. In surface mining, enormous open pits are dug to expose and remove coal-bearing rock. When the coal seams are exposed to air and water, toxic chemicals are formed that can impact local soils and waterways. Although modern oil-drilling techniques have minimized the environmental damage of removing oil from underground, hazards remain when oil is transported through pipelines or by tanker ships. Oil spills such as the 407,400 gallons (154,000 L) spilled from the Keystone Pipeline in South Dakota in November 2017 can kill thousands of animals, plants, and people, and have long-lasting harmful effects on fragile ecosystems. The website Vice News explained why this particular spill was so dangerous:

> *A viscous [thick] type of oil called diluted bitumen, or tar sands oil, flows through the Keystone Pipeline. Because it's so thick, leaks can be difficult to detect. If the oil does spill, it's often far more detrimental to sensitive water resources. Bitumen is also one of the dirtiest fuels in the world. Unlike conventional crude which can be pumped directly from the ground, water is required to separate the heavy, tar-like substance from the sand it's found in—a process that depletes and pollutes freshwater.*[11]

Oil spills can take several years and billions of dollars to clean up, and while cleanup efforts are taking place, people and animals must still deal with their polluted environment.

Serious harm to the environment also results when fossil fuels are burned to create energy. The exhaust from burning gasoline and diesel fuel in car engines and coal in factories and power plants creates a toxic mix of pollution. Once in the air, these pollutants, including carbon monoxide and nitrogen oxide, undergo chemical reactions to form the visible pollution known as smog. Smog can cause breathing problems such as asthma, and it contributes to birth defects. Smog-forming reactions occur when pollutants are blown through the air, so smog is often present in places far from its source. Smog also reduces visibility. In major cities such as Beijing, China, where there are many sources of pollution—including cars, factories, and coal-powered heating systems—visibility can fall as low as about 300 feet (91 meters), which can make it difficult for people to see where they are going. In natural environments such as national parks, where people often go to look at the scenery, pollution can interfere with the view. According to the National Park Service (NPS), "Air pollution can create a white or brown haze that affects how far we can see. Air pollution also affects how well we are able to see the colors, forms, and textures of a scenic vista."[12]

Burning fossil fuels creates smog, especially in big cities. It can make the air hard to breathe and cause multiple health problems.

Another far-reaching consequence of car and factory exhaust is acid rain. Acid rain is formed when sulfur and nitrogen released from burning coal and gasoline react with oxygen and water in the atmosphere. These chemical compounds are carried long distances on air currents, eventually falling to Earth in rain and snow. Acid rain changes the pH level in lakes and streams, killing microorganisms, invertebrates, and fish. Many plants are also affected by acid rain. Huge tracts of forests in Canada, the northeastern United States, Europe, and Asia have been damaged or killed by acid rain.

Effects on the Climate

Whereas smog and the effects of acid rain are easily visible, the impact of another, potentially much greater, threat is much more difficult to see—so difficult, in fact, that some people do not even believe it exists. Fossil fuels are primarily composed of the decayed remains of ancient plants and animals. Those remains hold the carbon that living plants once absorbed from the carbon dioxide in the atmosphere. When fossil fuels are burned, the carbon in the fuel combines with oxygen, forming carbon dioxide that is released back into the atmosphere. Some of this carbon dioxide is recaptured by growing plants as they absorb carbon dioxide through photosynthesis. However, carbon stores in fossil fuels were deposited over millions of years but are being released at a much more rapid rate.

For example, in 1997, ecologist Jeffrey S. Dukes of Purdue University calculated that the entire human population burned the equivalent of 400 times the total plant material grown that year. Releasing such large quantities of carbon dioxide through the burning of fossil fuels has resulted in a gradual rise in the levels of carbon dioxide in the atmosphere. In the 200 years between the beginning of the Industrial Revolution and 2005, the concentration of carbon dioxide in the atmosphere nearly tripled. In fact, data from the Mauna Loa Observatory in Hawai'i indicates that current carbon dioxide levels are at about 410 parts per million (ppm)—the highest ever recorded.

Carbon dioxide is considered a greenhouse gas, which means it is responsible for trapping the earth's heat in the

Denying the Problem

Some people do not believe that climate change exists or, if it exists, that it is truly a problem. There are several different reasons for this. Some people believe the data has been inaccurately reported, especially when it conflicts with what they observe. For instance, many people say global warming is not real because it still snows in some parts of the world. This represents a misunderstanding of the difference between weather, which changes daily, and climate, which is the general trend of an area's weather over a long period of time. Others do not believe global warming is caused by human activity, pointing to natural changes in the earth's climate that have taken place in the past. Scientists have addressed this by noting that the changes in the past took place much more slowly than they are happening today. Still others choose to believe what makes them the most comfortable: If global warming and climate change are not real threats, they do not have to worry or make any changes to their lifestyle.

atmosphere the way a greenhouse traps heat. Scientists believe that with higher atmospheric levels of carbon dioxide, less heat will be able to escape into space, resulting in a gradual rise in the earth's temperature—a phenomenon known as global warming.

Environmentalist and former vice president Al Gore has long sounded the alarm about the disastrous consequences that may occur from climate change, writing, "'Global warming' is the name it was given a long time ago. But it should be understood for what it is: a planetary emergency that now threatens human civilization on multiple fronts."[13] However, even some scientists doubt such a doomsday scenario.

Since 1880—the year when people began to keep detailed temperature records—the earth's average temperature has risen by 2 degrees Fahrenheit (1.1 degrees Celsius). It is extremely difficult to predict how much it will increase in the future. If carbon dioxide emissions continue to climb with the ever-increasing burning of fossil fuels, the rate of temperature increase is expected to accelerate. Even if societies are able to decrease their emissions, though, the temperature will continue to rise for a while. This is known as "committed warming," and it would happen because the carbon dioxide that is already in the air will not instantly disappear even if no new gases are added. Scientists say that if humans stopped all carbon emissions today, committed warming would still cause the temperature to rise to 2.3 degrees Fahrenheit (1.3 degrees Celsius) beyond pre-industrial temperatures by the year 2100. However, this type of quick stop is impossible, so people will continue adding greenhouse gases to the air for at least several more years—perhaps more, if governments do not work to place restrictions on emissions and search for alternative energy solutions.

Climate change is creating serious problems for plants and animals as well as humans.

Two degrees may seem like a small increase in temperature, but environmental changes have already been noted. Polar ice and glaciers worldwide are melting, birds and butterflies are shifting their ranges northward, and hurricanes are gaining strength. "There is no debate among any statured [respected] scientists working on this issue about the larger trends of what is happening to the climate,"[14] said James McCarthy, co-chair of the Intergovernmental Panel on Climate Change (IPCC). Many experts believe these changes are just a small taste of the environmental storm brewing. If temperatures continue to increase at an accelerated rate, entire ice sheets in Greenland and Antarctica may melt, raising global sea levels enough to swamp large parts of coastal Florida, many Pacific islands, and parts of Southeast Asia. Entire ecosystems may be altered as plants and animals either adapt or die out in response to the changing climate.

A DANGEROUS ADDICTION

"We just can't seem to break our addiction to the kinds of fuel that will bring back a climate last seen by the dinosaurs, a climate that will drown our coastal cities and wreak havoc on the environment and our ability to feed ourselves. All the while, the glorious sun pours immaculate free energy down upon us, more than we will ever need … The dinosaurs never saw that asteroid coming. What's our excuse?"

–Neil deGrasse Tyson, astrophysicist

Neil deGrasse Tyson, "The Lost Worlds of Planet Earth," *Cosmos: A Spacetime Odyssey*, directed by Brannon Braga, Ann Druyan, and Bill Pope, May 4, 2014 (Los Angeles, CA: 20th Century Fox, 2014), DVD.

Problems with Oil Dependence

Although some disagreement exists about the potential impact of global warming, there is little argument that America's appetite for oil has serious consequences. In the past, the foremost concern to most people—politicians and the general public alike—was that America could not produce nearly enough oil to fuel its own energy needs. In fact, the United States produced only about half the oil it consumed. As of 2016, however, that number has increased to about 75 percent. Some people see this as a good thing because it has lessened the country's reliance on importing foreign oil from countries such as Nigeria, Venezuela, and Iraq—regions that are home to ongoing political conflicts and governments that do not always share the views of the United States.

"Oil addiction is America's Achilles' heel," explained author Matthew Yeomans. "To remain dominant in the world, the U.S. must be sure that oil flows freely and consistently onto the global market."[15] Robert E. Ebel of the Center for Strategic and International Studies agreed, saying, "Oil fuels more than automobiles and airplanes. Oil fuels military power, national treasuries, and international politics."[16]

Since the 1980s, America has increased its military presence in the Middle East as well as in certain oil-producing countries in South America, other parts of Asia, and Africa. Author Michael T. Klare examined this trend in his book *Blood and Oil*:

> *Taken together, these developments lead to an inescapable conclusion: that the American military is being used more and more for the protection of overseas oil fields and the supply routes that connect them to the United States and its allies … Slowly but surely, the U.S. military is being converted into a global oil-protection service.*[17]

Top military officials have admitted the truth of this. In 2007, Chuck Hagel, who was a senator at the time and served as the Secretary of Defense from 2013 to 2015, said of the Iraq War, "People say we're not fighting for oil. Of course we are."[18] Some people do not approve of this use of the military, and others believe the United States should not have to rely on governments whose actions it may not support. For this reason, many are happy with the country's growing energy independence.

However, others believe this newfound self-sufficiency comes at too high a price. The United States has reduced its reliance on foreign oil through offshore drilling and fracking, both of which are controversial. Offshore drilling is the process of extracting oil that can be found under the ocean off the coast of the United States. Fracking is "the process of drilling down into the earth before a high-pressure water mixture is directed at the rock to release the gas inside."[19] Other countries, such as the United Kingdom (UK), have also investigated these methods because of the boost they have given the U.S. economy. According to the BBC,

> *In the US [fracking] has significantly boosted domestic oil production and driven down gas prices. It is estimated to have offered gas security to the US and Canada for about 100 years, and has presented an opportunity to generate electricity at half the CO2 emissions of coal.*[20]

Similarly, supporters of offshore drilling say it would also lower gas prices and decrease U.S. reliance on foreign oil. However, many people oppose fracking and offshore drilling because of the possible dangers they pose to the environment as well as people's lives and health. For instance, both methods have the potential to leak dangerous chemicals into the environment and contaminate an area's water supply, killing animals, plants, and people. Additionally, some experts have expressed concerns that breaking apart rock this way could cause some earthquake activity; in fact, according to the British newspaper *The Guardian*, the Netherlands has halted fracking after several large earthquakes damaged homes and businesses near the Groningen field, one of Europe's largest sources of natural gas. Eric Weibes, the minister for economic affairs, told the gas extraction companies that they needed to start pursuing alternative forms of energy.

Fracking and offshore drilling have also had unintended negative effects on animals in the surrounding areas: "ExxonMobil recently had to suspend exploration efforts near Madagascar after more than 100 whales beached themselves"[21] due to the seismic waves used to locate the oil under the rock. These waves confuse whales, so instead of moving deeper into the ocean, they move toward the beach, where they get stuck and die.

People who support the economic benefits of fracking and offshore drilling often point to the improved technology and government rules that they say

Many people are strongly against fracking. Although it provides a source of energy, many people believe it is not worth the risk to their health.

make these processes safe. Although it is true that offshore drilling has had a high safety record since 1975, opponents have pointed out the increased potential for accidents since President Donald Trump's administration has repealed President Barack Obama's safety regulations. Additionally, even one accident can have devastating consequences for the area it affects. Activists say that a more modern solution based on efficiency and renewable resources is necessary rather than continuing to contribute to climate change with fossil fuels.

Rethinking Transportation

The world is increasingly interconnected; people rely on cars, trains, buses, and airplanes to help them conduct business, visit family and friends, and see new places. Trucks, boats, and other large vehicles also transport goods from one place to another, which is why it is possible for someone in Montana to buy fresh Florida oranges or Mexican avocados at the grocery store. All of these forms of transportation are necessary in today's world, but a lot of oil is needed to power them; according to the U.S. Energy Information Administration (EIA), as of a 2017 report, transportation accounts for 29 percent of the energy used in the United States. For this reason, finding new ways to power vehicles is crucial.

Reliance on Cars

According to the World Economic Forum, the top 10 countries with the most registered vehicles on the road are mostly European. For example, Finland is the world leader with more than one vehicle for every person who lives there. The United States ranks fourth in the world, but according to the research group Shrink That Footprint, Americans drive more miles per year than people in any other country, even compared to countries that are as large as or larger than the United States. For instance, Americans drive 40 percent more miles each year than Canadians and 60 percent more than Australians. There are several reasons for Americans' dependence on cars:

- Many European countries have created policies that encourage alternate forms of transportation, such as walking or bicycling.

- Mass production in the 1930s made cars cheaper in the United States than in other countries, which encouraged more people to buy them.
- The increased amount of cars on the road led U.S. road planners to create road systems better suited to driving; for instance, roads are often much wider in the United States than in Europe, and most U.S. cities are connected by major roadways, while it may be difficult or take a long time to move between cities in other countries.
- Transportation taxes tend to be higher in other countries, while public transportation tends to be cheaper and more readily available.

Despite rising gasoline prices, fears of dependence on foreign sources of oil, and a growing awareness of the environmental harm that cars cause, American consumers have so far proven unable to make significant changes in their driving habits. Cars offer a lifestyle of easy mobility and convenience that Americans have grown to depend on. Options that would reduce Americans' dependence on cars, such as comprehensive public transportation systems, have not been well developed. Clearly, cars are here to stay. "The bottom line is we can't change America's love affair with the automobile, but we can change the automobile,"[22] said Dan Becker, former director of the Sierra Club, which is a conservation group that promotes alternative vehicles among other energy alternatives.

Roads connect places in the United States where there is no public transportation, so in many places, cars are essential. Additionally, many Americans like to take scenic drives.

Reworking Current Cars

Some people believe that instead of creating cars that are not powered with fossil fuels, it makes more sense to simply improve the way current cars work. One relatively new feature in many cars is called stop/start technology, which helps cars go further on the amount of fuel their tank can hold. Cars with this technology automatically shut off when the brake pedal is pressed for a few seconds. As soon as the driver takes their foot off the brake, it starts up again. This means the car is not using fuel at red lights or in traffic jams. Drivers say the transition is so smooth that they barely notice it happening.

Another measure that has been put in place is emission limits, which are set by the government. They dictate how much carbon dioxide a car can emit. Cars that fail an emissions test cannot be registered, which means it is illegal for someone to drive them. Although this is one step toward more eco-friendly cars, many people believe it is not far enough. Additionally, Volkswagen came under fire in 2016 when it was revealed that it had been lying about its cars' fuel emissions by falsifying the results of their emissions tests. According to *Fortune* magazine, this had been going on for about 10 years, resulting in about 10.5 million Volkswagen cars around the world that were emitting up to 40 times the legal amount of greenhouse gas emissions. This shows that relying on companies to follow environmental laws is not always advisable, since they have an incentive to do things that will save them the most money. In Volkswagen's case, this meant making a less environmentally friendly engine that cost them less money to produce.

Powering Cars with Electricity

Using alternative power sources for vehicles is not a totally new idea. In the 1970s, as pollution increased and gasoline shortages began, the electric car was introduced as the first alternative to traditional cars. Electric cars do not use gasoline or diesel fuel; instead, they use electric motors powered by current from onboard batteries. Once considered the most promising new vehicle technology, the electric car lost most of its early appeal because of its limited range. The electric car's weakness is its battery. Most can provide enough electricity to power an electric car's travel for a maximum of only 80 miles (128 km), after which time they must be plugged in for 4 to 8 hours to recharge. This short range prevents the electric car from being a practical alternative for most American drivers, who often make longer trips. While there are many gas stations around to help people refuel, there are very few electric car recharging stations, and it takes much longer to charge a car than to fill up its tank with gas.

Despite these downsides, the electric car does have significant benefits: It costs less to maintain, costs less than most conventional cars to buy, and lasts longer than a conventional car since it has fewer moving parts to wear out. The electricity cost of charging the car is lower than the gasoline costs for a conventional car. Most importantly, since it does not use oil or gasoline, an electric car creates no pollution when running. The California Air Resources Board has studied electric cars and concluded that even when factoring in pollution from power plants, electric vehicles reduce pollution by more than 90 percent compared to the newest and cleanest conventional vehicles.

The first commercial electric vehicle (EV) was the EV1, developed by General Motors in 1996. However, the car was slow to catch on. People were nervous about having to plug in the cars, and some were unhappy with the car's performance. Eventually, General Motors discontinued the EV1. This decision angered many electric car supporters, who thought it was the end of the technology, but other companies stepped up to create their own EVs. Since 2010, the best-selling EV has been the Nissan Leaf, according to *Popular Mechanics*.

New battery technology promises to improve the range of electric cars, overcoming the major stumbling block to developing broad appeal to consumers. New electric car entrepreneurs are trying to help make the cars popular and affordable. In 2017, for instance, Samsung released a battery that charges to 80 percent within 20 minutes. This charge can take the car up to

The EV1, shown here, was the first electric car available for people to buy. However, it did not sell well because of its limited range, among other reasons.

310 miles (500 km)—further than most batteries can take a car on a full charge. A full charge on this new battery increases the car's range to about 370 miles (600 km). This type of battery could be a key to making electric cars more popular, as it would eliminate the long wait time at rest stops and reduce fears about how far the driver can go. Other companies are trying to create similar batteries.

The cost per car varies depending on the brand; a new Tesla costs about $70,000, while the Leaf costs about $29,000 (or $34,000 with the new battery pack). In comparison, the average new conventional car costs about $32,000. Audi, BMW, and other companies have also developed or are in the process of developing electric cars. Although electric cars are still not common in most places, they are gaining popularity; for

A SLOW PROCESS

"Electric cars are not going to take the market by storm, but it's going to be a gradual improvement."

–Carlos Ghosn, chairman and chief executive officer of Renault-Nissan-Mitsubishi

Quoted in Dave Lee, "Nissan-Renault Head Carlos Ghosn's Zero Emission Goal," BBC, February 9, 2010. news.bbc.co.uk/2/hi/business/8501348.stm.

example, some news outlets have reported that Norway is debating a law that would ban all cars that run on fossil fuels by 2025. As Wes Raynal of *AutoWeek* pointed out, "no one knows what's coming in terms of fuel costs, customer wishes, fast-charging infrastructure, battery development costs, electricity generation and a gazillion other factors I can't think of right now."[23]

Combining Electricity and Conventional Fuel

Because of the problems that must be overcome in the production of electric cars, most car manufacturers are looking at other alternatives. Much research focuses on boosting the fuel efficiency of cars—finding ways to drive longer distances using less gas.

Car manufacturers have been able to achieve a higher level of fuel efficiency through the development of the hybrid car. The hybrid car is essentially a cross between a gasoline- or diesel-powered car and an electric car, combining the best features of both types of vehicles. Hybrid cars run on both a gasoline or diesel engine and an electric motor powered by batteries. The batteries in a hybrid car never need to be plugged in. Instead, the gas engine runs a generator that keeps the batteries charged. Because the electric motor helps power the car, the gas engine of a hybrid is smaller than that of conventional cars and requires less fuel. The hybrid is controlled by a computer that

Hybrid car engines such as the one shown here have fuel lines as well as a large battery.

automatically coordinates the operation of the engine and electric motor. At low speeds, the hybrid operates on the electric engine alone. Once a higher speed is reached, the gas engine takes over. Because the hybrid is under automatic control, drivers notice little difference from a conventional car.

The hybrid's use of an electric motor to boost power allows drivers to save a lot of money on fuel. Many hybrids can travel twice as far on a gallon of gas as a conventional car. The Toyota Prius Two Eco, for example, averages 56 miles per gallon (24 km/l), and the BMW i3 94 Ah averages 118 miles per gallon (50 km/l). In comparison, the most fuel-efficient conventional car averages about 40 miles per gallon (17 km/l). Because they use less gas, hybrids cost less to fill up, although some people have noted that when gas prices are low, the gas savings do not offset the upfront price. According to some experts, when gas prices are low and the cost of the car is high, it can take years to make up the price difference. However, for many people, the true benefit is that using less gas creates less pollution.

Since the introduction of the first commercially available hybrid car in the United States in 1999, sales have doubled every year, supported in part by tax incentives available to those who purchase hybrids. The increasing popularity of hybrids has the potential eventually to make a significant impact on the amount of oil used to run vehicles in the United States. However, most experts caution that because they still require gas to operate, hybrids are only a temporary solution to declining oil supplies.

Concerns About Electric and Hybrid Cars

Although hybrid and electric cars are considered better for the environment because they use less gas and oil, some people have noted that certain problems will arise if they become widespread. For instance, concerns have been raised about what will happen to the car's lithium-ion battery when it can no longer be recharged and must be replaced. Lithium-ion batteries—the most common kind currently available—can be recycled, but many people do not bother to take the steps to do so; according to *The Guardian*, only about 5 percent of batteries in the European Union (EU) are currently recycled. This can be harmful to the environment: "Not only do the batteries carry a risk of giving off toxic gases if damaged, but core ingredients such as lithium and cobalt are finite [limited] and extraction can lead to water pollution and depletion among other environmental consequences."[24]

In the case of car batteries, some countries have passed laws requiring car companies to be responsible for collecting and recycling them. When small lithium-ion batteries such as AA or AAA batteries wear out, many people put them in a drawer or throw them in the garbage, but car batteries would be too big to store at home, and many landfills would not accept them. Another option, which Nissan is investigating, is for car batteries to be repurposed to provide energy at home rather than in the car. Other entrepreneurs are investigating different recycling techniques that would be less expensive and more environmentally friendly.

A bigger question is how long and how ethically batteries can be produced. According to the *New York Times*, there are several concerns in this regard:

Some countries where cobalt is found naturally have been accused of human rights violations, such as employing children as cobalt miners.

Three-quarters of the world's reserves of lithium … are in China and Chile, according to a United States Geological Survey. As demand surges, China could deploy its natural resources as a diplomatic cudgel [threat] the same way that Saudi Arabia uses oil …

As for cobalt, it comes primarily from the Democratic Republic of Congo, one of the world's most war-torn and unstable countries. Illegal mining operations there have been accused of using child labor.[25]

Mining for the materials needed to create batteries has destroyed certain areas—often in impoverished or conflicted regions, where the local residents do not benefit from or are actively harmed by the mining and have little or no say in whether it takes place. It has been said that "increasing our dependence on electric and hybrid cars may mean trading the conflicts in the Middle East for another set of problems,"[26] including high prices due to demand and scarcity, battles for mining territory, doing business with countries whose business practices violate human rights, and causing environmental harm. Additionally, demand for other types of batteries—such as the ones that power laptops, cell phones, and other electronic devices—will likely remain stable or increase, placing an extra strain on resources.

To try to address these issues, some companies, such as Tesla, have pledged to use only ethically sourced materials from North America. However, since the supply of cobalt and lithium is limited in North America, it is uncertain how long this can last. Recycling can also stretch the supply of these materials; according to the Union of Concerned Scientists, "Cobalt … (as opposed to gasoline) is fully recyclable and roughly 15 percent of U.S. cobalt consumption is from recycled scrap today."[27] Another solution in the works is changing battery technology to rely less on cobalt or lithium. For instance, some batteries are powered by magnesium, sodium, iron, or sulfur. Since 2012, products presented at the annual Battery Symposium in Japan have moved away from lithium-ion power toward other options such as lithium-air or non-lithium power. The technology is still evolving, but if ethically sourced, sustainable batteries can be created, electric and hybrid vehicles may become a better option in the future.

Alternative Fuel Options

Although hybrid cars are currently the most widely available and popular approach to curbing the use of oil for powering vehicles, other solutions are quickly gaining prominence. One of the most promising is the use of biofuels, or fuels made from plants. Plants capture the energy of the sun through photosynthesis and store this energy in their tissues, such as their leaves, stems, and wood. When plants are used for fuel, they release the stored energy in the same way fossil fuels do. However, to make biofuels, plants are not just burned to create energy. Instead, biofuels are made through a chemical process that transforms the energy in plant tissues into a liquid fuel that can be used much like gasoline or diesel fuel.

There are two main types of biofuels: biodiesel and ethanol. Biodiesel is made from the vegetable oil found in plants such as soybeans and peanuts. It can also be made from animal fats and cooking oil recycled from restaurants. Several different proportions of biodiesel exist; biodiesel mixed with regular diesel fuel in the proportion of 20 percent biodiesel to 80 percent diesel (known as B20) is the most common type of biodiesel used in the United States. It can generally be used in any standard diesel engine. The pure form of biodiesel (known as B100, or 100 percent biodiesel) is rarely used in engines; it is most commonly used to create other mixtures of biodiesel, according to the U.S. Department of Energy. This is because the higher the concentration of biodiesel rises above 20 percent, the lower the amount of energy the fuel gives off, which means people have to stop to refuel more often.

ON A MISSION

"Our mission is to make a gas that is so cheap and plentiful that consumers don't even have to know it's not made from fossil fuels."

—Martin Tobias, entrepreneur and investor in biodiesel

Quoted in Julie Halpert, "Ten Eco-Friendly Companies," *Newsweek*, November 20, 2005. www.newsweek.com/ten-eco-friendly-companies-115177.

Biodiesel can be recycled from restaurants' leftover oil and grease.

Instead of being made from oil-rich plants, ethanol is made from plants high in sugar, such as corn, sugar cane, and sugar beets. Through the process of fermentation, the sugar in the plants is transformed into ethanol, a type of alcohol. One type of ethanol, called cellulosic ethanol, is generally made from corn husks, wood chips, and other inedible plant parts. This type produces even fewer emissions than ethanol with a base made of edible plants.

Like biodiesel, ethanol is mixed with unleaded gasoline in different amounts; according to the Department of Energy, more than 97 percent of gasoline in the United States contains at least some ethanol as a pollution-reducing measure. Any engine can run E10, which is 10 percent ethanol to 90 percent gasoline; this is the type most commonly found in the United States. Cars that were made after 2001 can also use E15. Vehicles that run on E85—which, contrary to the straightforward abbreviations of the other fuel types, is 83 percent ethanol to 17 percent gasoline—are called flexible fuel, or flex-fuel, vehicles because they can also run on unleaded gasoline.

Both ethanol and biodiesel can be transported and sold in a similar manner as standard vehicle fuel. "Biofuels are the easiest fuels to slot into the existing fuel system,"[28] said Michael Pacheco, former director of the National Bioenergy Center. Biofuels tend to be cheaper at the pump than gasoline, and when burned in cars, they produce significantly less carbon dioxide and pollutants that contribute to smog and acid rain. They are nontoxic and will break down more easily in the environment if they are spilled. Perhaps most importantly, biofuels are made up of mostly renewable plant resources. In their pure forms, biofuels will be available even when fossil fuels are used up.

The Myth of Chemtrails

As a plane flies, it leaves behind a familiar sight: thin, white, cloud-like streaks that remain in the sky for a while after the plane passes. These are called contrails, and they are formed when jet engines emit hot, humid air high up in the sky, where temperatures can be as cold as -40 degrees Fahrenheit (-40 degrees Celsius). At this temperature, the water vapor in the contrail condenses, forming a cirrus cloud. Although not harmful to people on the ground or in the plane, contrails have been found to have a negative effect on the environment by trapping sunlight in the atmosphere, reflecting it back to the ground instead of letting it escape into space. Switching to biodiesel could limit the amount of contrails in addition to reducing carbon emissions.

A common myth is that contrails contain dangerous chemicals the government is using to experiment on citizens, leading to the nickname "chemtrails." People have many different conspiracy theories about the purpose of chemtrails; for instance, that they are being used to control citizens' minds, prevent them from being able to have children, or make them sick. Scientists say such experiments would not work, as contrails are too high in the sky to make a difference to those on the ground. Additionally, even if they did work, it would be difficult for the government to observe the effects on the population, as they would not know for sure who had been affected; according to experts, "the chemtrails would be released so high in the atmosphere that unpredictable winds would move them around significantly, making such experiments worthless and unreliable."[1]

1. Jane McGrath and John Perritano, "What Are Chemtrails, and Should You Be Scared of Them?," How Stuff Works, March 7, 2011. science.howstuffworks.com/transport/flight/modern/what-are-chemtrails1.htm.

Biofuels are also improving air travel. Many people travel for fun or business, and products such as food and mail are also shipped through the air, so flights account for a large part of transportation. According to the Federal Aviation Administration (FAA), each day there are more than 26,000 scheduled passenger flights in the United States alone. Aviation fuel, also known as jet fuel, produces carbon emissions just as gasoline does, so it is also contributing to air pollution. In 2017, the National Aeronautics and Space Administration (NASA) published a study in the journal *Nature* about the effects of adding biofuel to aviation fuel. The study found that blending 50 percent of each type of fuel reduces emissions by 50 to 70 percent, which makes experts hopeful that this mixture will lessen airplanes' negative effects on the environment.

Controversy over Biofuels

Some people believe biofuels are the answer—or at least one answer—to the problems of fossil fuels, but others strongly disagree. One thing that makes them popular is that the cost of filling up a gas tank with biofuel is equal to or less than filling it with gasoline. According to the Department of Energy, in October 2017, the average price of a gallon of unleaded gasoline in the United States was $2.49, compared to $2.10 for E85. B20 cost slightly more at $2.68, but it cost less than traditional diesel fuel, which cost $2.76. However, those willing to buy biofuels face the challenge of finding gas stations that carry those fuels. In 2005, for example, there were only 600 E85 fueling stations in 30 states. As of 2018, there are more than 3,000 in 43 states, but this is still not considered a large amount, considering that there are around 150,000 gas stations overall in the United States. Opponents of biofuels point out that they are not as energy-efficient as gasoline; some say this means people are not actually saving as much money as they think they are because they must fill up more often.

Another issue opponents point out is where the materials to make biofuels come from. Because they are made from food, biofuel production has a direct impact as well as several indirect effects on food supply. According to Chris Smith, the cofounder of a science podcast called *The Naked Scientists*,

> *In the case of US corn, for example, using the crop to produce fuel means that the price of corn to eat goes up. So what, you might ask, I can live without sweetcorn? But corn is widely used as an animal feed, so the price of beef goes up. Livestock farmers then look for alternative feed and the price of other grains—wheat, barley or rice—goes up. So using corn for fuel makes the meat and the bread in your burger more expensive!*[29]

This increase in prices may not mean much to most people in developed countries, "but in the developing world higher food prices can be catastrophic. Food riots and political instability can result."[30] Riots have already occurred in Mexico over the rising cost of corn tortillas. In early 2018, the National Tortilla Council warned that it could raise prices up to 20 percent, which triggered more outrage and threats of protests on social media. Mexican economic officials called the price raises unjustified and agreed to work with the Tortilla Council to keep costs down.

Concerns have also been raised about the long-term impact on the environment of growing the crops needed to create biofuels and turning them into usable fuel. For example, soy is the main crop used to create biodiesel, but it takes "five times more land than ethanol to produce the equivalent amount of biofuel energy."[31] To supply the whole world with biofuels, rain forests and other ecosystems would have to be converted to farmland, which would have serious negative consequences on the

One problem with biofuel is that it takes huge amounts of farmland to produce enough to meet the demand for both food and fuel.

diversity of plant life and the animals that live in those areas as well as using huge amounts of fossil fuel–powered energy in the conversion process.

Additionally, although biodiesel does not emit carbon dioxide, it does emit other greenhouse gases that some experts say have a much higher impact on global warming. Furthermore, growing and harvesting the crops, powering the factories that turn the plants into fuel, and transporting the fuel require large amounts of energy, generally from fossil fuels. Estimates of the actual amount of energy gained by making biofuels range widely. One study showed that it takes two-thirds of 1 gallon (2.5 l) of oil to make 1 gallon (3.8 l) of ethanol from corn. Therefore, using a gallon of ethanol would save only one-third of 1 gallon (1.25 l) of gas. "There are many in the energy community who believe that the amount in the oil that you use to produce a gallon of ethanol is about the same as you get out of it once you've made it,"[32] said Ryan Wiser, a scientist at Lawrence Berkeley National Laboratory. Since it is less energy-efficient than gasoline, more than a gallon of biofuel must be produced to provide the same amount of power as a gallon of gasoline, which means biofuel may actually end up having a negative effect on the environment instead of a positive one. For this reason, some people believe biofuels should remain an optional alternative to fossil fuels rather than becoming the main fuel source; in 2014, for instance, the United Nations (UN) warned that widespread use of biofuels could actually increase global warming rather than decrease it.

Turning to Hydrogen

Hydrogen is another alternative vehicle fuel technology that is generating corporate and government interest and research funding. Hydrogen is the most abundant element in the universe, but on Earth, it rarely exists in its free state. Instead, it binds to other elements such as oxygen and carbon to form water, plant material, and fossil fuels. Instead of being an energy source itself, such as oil or coal, hydrogen is an energy carrier in the same manner as electricity. In other words, to provide energy, hydrogen must first be produced using an energy source to free it from its bonds with other elements.

AN ETHICAL PROBLEM

"For the U.S. and the world, 48 million Americans live in poverty, and over 80% of the globe is undeveloped, so the rising competition between 'fuel and food' is a moral issue."

–Jude Clemente, *Forbes* contributor

Jude Clemente, "Why Biofuels Can't Replace Oil," *Forbes*, June 17, 2015. www.forbes.com/sites/judeclemente/2015/06/17/why-biofuels-cant-replace-oil/#3cf1a85ff60f

Hydrogen powers a car by creating electricity within a special onboard device called a fuel cell. A chemical reaction takes place as hydrogen enters the fuel cell, splitting the electrons off the hydrogen molecules to produce electricity that powers the car's motor. Fuel cell–powered cars produce no pollution or greenhouse gases; only a small amount of water emerges from the tailpipe as a result of the chemical reaction.

Hydrogen may seem like an environmentalist's dream solution to the problems of fossil fuels. However, developing hydrogen as a widespread vehicle fuel has many stumbling blocks that must be resolved before this dream can become reality. Perhaps the most significant challenge lies with the production of the hydrogen itself. Currently, the raw materials used to produce hydrogen are primarily natural gas and coal. Extracting hydrogen from these materials requires a great deal of energy and releases carbon. In fact, the amount of energy required to produce hydrogen is about the same as the amount of energy hydrogen produces to propel a car. So although a hydrogen-powered car may produce no pollution itself, the entire process still requires fossil fuels and creates harmful emissions.

To solve these problems, researchers are focusing on ways to produce hydrogen from water through a process called electrolysis. Although it eliminates the need for fossil fuels for raw material, electrolysis is a complex and expensive process that also requires a great deal of energy. If this energy comes from renewable sources such as solar or wind power, the process is much better for the environment but still expensive, as are other methods, such as using ethanol to create a reaction with high-temperature steam. The cost of production and transportation makes hydrogen fuel more expensive than gasoline, and there is currently no good way to transport it more than 700 miles (1,126.5 km) from where it was produced. These are important considerations when looking at alternative energy sources because if the product is too expensive, few people will be able to afford it and it cannot be widely adopted even if there are no other problems with it.

Cars that are powered by hydrogen must carry the fuel in special tanks, such as the one shown here.

Producing hydrogen is only the first step. To be an effective vehicle fuel, it must be easy for consumers to purchase and store in their cars. Hydrogen is most commonly used in its gas form. However, hydrogen gas takes up a great deal more space than its energy equivalent in gasoline. This makes it difficult to store a large amount in the small space of a car's gas tank. To overcome this obstacle, hydrogen gas can be compressed and stored in strong, high-pressure tanks, which can be refilled at special refueling stations in about 5 minutes, although there are few of these refueling stations currently available in the United States; as of 2018, there are only 25 fueling stations, and they are all in California. However, these tanks pose a risk of explosion if punctured. The smaller the tank, the safer it is, but this introduces the same problem as electric batteries: a range that is limited to about 100 miles, or 160.9 km (a larger hydrogen tank is riskier but can take a car up to 300 miles [482.8 km]). Hydrogen can also be stored as a liquid but needs to be cooled to -423 degrees Fahrenheit (-253 degrees Celsius). If it is not used quickly after it arrives, it evaporates and becomes useless. Because hydrogen is highly flammable, further research is required to develop a storage system that will be both safe and cost-effective. Further research is also necessary to extend the life of fuel cells, which currently wear out quickly and are difficult to replace.

FINDING BETTER ALTERNATIVES

"The Stone Age did not end because we ran out of stones, and the oil age will not end because we run out of oil."

–Don Huberts, CEO of Shell Hydrogen

Quoted in Christine Woodside, *The Homeowner's Guide to Energy Independence*. Guilford, CT: Lyons, 2006, p. 64.

Despite the challenges, many alternative energy experts have expressed the view that hydrogen will play a large role in the future of global energy use, and they are slowly gaining popularity, although electric and hybrid cars remain far more popular. Price remains a significant barrier. For example, in the early 2000s, Iceland stated that it would be moving from fossil fuels to hydrogen. However, those plans never progressed; according to the Worldwatch Institute in 2006, "it seems that Iceland is passively relying on others to shoulder the initiative and the critical funding of research, while merely offering the country as a staging ground."[33] In other words, while Iceland wanted hydrogen technology, it did not have enough money to put it into place. In 2009, Iceland's leaders admitted that the country's financial crisis had set the hydrogen plan back at least 10 years. In February 2017, a Norwegian company named NEL Hydrogen signed an agreement with an Icelandic company called Skeljungur to expand Iceland's fleet of hydrogen-powered buses from three to ten, so it is possible that the plan is getting back on track.

Perhaps the best indicator that a transition to hydrogen may be underway is the fact that many oil companies and automakers have committed billions of dollars to this developing technology. In addition to the NEL and Skeljungur agreement, in 2017, Shell, BMW, and 11 other companies "formed a council to accelerate the adoption of hydrogen as a transportation fuel."[34] Seth Dunn, former researcher with the Worldwatch Institute stated the need for urgency: "Countries that focus their efforts on producing oil until the resource is gone will be left behind in the rush for tomorrow's prize."[35] Whether the ultimate "prize" will be hydrogen fuel cells, biofuels, or hybrid or electric cars remains to be seen.

CHAPTER 3

The Power of the Sun and Wind

Solar and wind power are two of the most commonly discussed forms of alternative energy. The sun generates an incredible amount of energy; according to *The Old Farmer's Almanac*, "In just one second, the Sun produces about 118 times as much energy as the United States used in the year 2003!"[36] Most of this energy is released into space, but the amount that does find its way to Earth keeps life going. Through the use of solar panels, it can also be harnessed and stored to power things that are currently running on fossil fuels. The sun's energy is renewable and releases no harmful gases or chemicals.

WIND IS TODAY'S TECHNOLOGY

"Wind works. It's reliable. It's economical. It makes environmental sense. And it's here now. Wind machines are not tomorrow's technology."

–Paul Gipe, author of *Wind Power: Renewable Energy for Home, Farm, and Business*

Paul Gipe, *Wind Power: Renewable Energy for Home, Farm, and Business*. White River Junction, VT: Chelsea Green, 2004, p. 1.

Solar power has been steadily gaining popularity; within a decade, it has gone from providing less than 0.1 percent of the world's electricity needs to more than 1 percent. Some countries use more solar power than others; for example, in Germany, solar power accounts for more than 7 percent of the country's total energy supply. As advances in solar technology continue to be made, these percentages are projected to grow, providing more energy and more jobs for workers.

Windmills have been used for hundreds of years to generate energy, although traditional windmills look much different than the ones designed today. They are particularly associated with the Netherlands in Europe (shown here), where they are a tourist attraction as well as a source of energy.

Wind power is also being used by many countries. In some respects, wind is actually a form of solar energy. As the sun shines down, it heats the atmosphere and the earth's surface. However, variation in land formations, vegetation, and bodies of water causes uneven heating around the planet. Air in some areas, such as at the equator, warms quickly. As this warm air expands and rises, cooler air rushes in to replace it. The warmed air eventually cools and sinks back to the earth. This movement of warm and cool air combined with the rotational force of the earth creates wind.

People have been harnessing energy from the wind for thousands of years. Long before the discovery of coal or oil, sailing vessels traveled the world powered only by the wind in their sails. Windmills used for pumping water and grinding grain have been a common sight in the countryside of Europe for centuries. In these cases, wind works directly by pushing ships and powering wheels to move water and grind grain. The energy that results stays in the location where it is generated. More recent technology, however, has allowed the wind's energy to be used indirectly by converting it into electricity that can be transmitted long distances and used for many different purposes.

Methods of Harnessing Solar Power

There are currently three ways to use the energy from the sun: passive solar technologies, solar thermal systems, and photovoltaic (PV) cells. Passive systems rely on building techniques and materials to save energy from the sun without using any mechanical means. Examples of passive solar energy use include greenhouses and homes designed with windows facing south to take maximum advantage of the sun's rays. Because buildings consume a great deal of energy for heating and cooling, these methods are valuable for improving a building's energy efficiency and reducing its fossil fuel needs. However, passive solar systems cannot transform the sun's energy into other useful forms.

Instead of simply saving the sun's energy, solar thermal systems use mechanical methods to collect, control, and store the flow of energy from sunlight. The most basic thermal systems consist of panels mounted on a building's roof. Sunlight heats water or other fluid running through pipes within the panels. This hot water is then pumped through a device called a heat exchanger that transfers the heat from the pipes to water in a storage tank. The heated water can be used for bathing or to warm a building's floor through radiant heating.

Solar power can bring benefits to people who live in places without a good electricity grid. Shown here are solar hot water heaters on the roofs of houses in Alexandra, South Africa.

Because these technologies are relatively easy and affordable to install, home solar hot water systems are in wide use around the world. Since water heaters represent up to 25 percent of the energy used by a home, a widespread shift to solar water heating systems has the potential to give people significant money and energy savings. Several countries, including Israel and Spain, now require solar water heaters to be installed in all new buildings. However, small-scale solar thermal systems provide only very localized and limited benefits from the sun's energy. That is, the heat in the water cannot be transported far from where it is gathered, and it cannot be used as energy to provide power for other uses.

Transferring Solar Power

Instead of simply transporting the sun's heat, larger, more complex solar thermal systems are used to convert solar energy into electricity that can be transmitted to multiple homes and businesses. Solar power plants are designed to capture as much energy from the sun as possible. In the United States, there are about 6,000 major solar plants in places such as California, Texas, North Carolina, and Minnesota, and more are being constructed all the time. Just as in some traditional power plants, an industrial solar thermal system heats water to create steam that is used to turn turbines, a process that generates electricity. With a solar plant, however, water is heated through the sun's energy instead of by burning coal or natural gas. To capture sunlight, these plants use hundreds of large, curved mirrors, called heliostats, specially programmed to rotate to track the movement of the sun. The mirrors reflect the sunlight in a concentrated beam aimed at a central tower. The intense heat that results boils water in the tower's tank, creating steam that drives the electricity-producing turbines.

Another kind of solar plant uses PV cells—similar to those seen on solar-powered calculators—to capture sunlight and convert it into electricity:

> *As light hits the solar panels, the solar radiation is converted into direct current electricity (DC). The direct current flows from the panels and is converted into alternating current (AC) used by local electric utilities. Finally, the electricity travels through transformers, and the voltage is boosted for delivery onto the transmission*

lines so local electric utilities can distribute the electricity to homes and businesses.[37]

PV cells are made from electricity-conducting materials called semiconductors—the same materials used to make computer chips. The most commonly used semiconductor is the element silicon. When light hits the PV cell, the silicon inside absorbs a portion of its energy. This transfer of energy knocks off some of the silicon atom's electrons, allowing them to flow freely and creating an electric current. This current can either be used directly or stored in batteries for later use.

Space-Based Solar Power

One proposed application of solar power seems more suited to science fiction than real life. Space-based solar power, or SBSP, would capture sunlight using huge PV arrays orbiting the earth. The energy generated would be converted to microwaves and sent down to Earth, where the microwaves would be captured by an antenna and transformed into usable electricity. This would allow solar power to be collected and used even when it is nighttime on Earth. It could also deliver energy to people in remote areas who are not connected to a power grid.

NASA first studied the concept of SBSP during the energy crisis of the 1970s, but technology did not support the idea at the time. In 1995, in light of advances in PV technology, NASA took another look at getting energy from space. NASA determined that within several decades, it would be possible to have a system in place that could generate small amounts of electricity. Many of the technologies needed to do this have already been invented, and more are being developed, although it will still be many years before they are put into place.

Solar cells have been used to make electricity since the 1950s. Early photovoltaic technology relied on large, bulky collectors that converted only 10 to 15 percent of the sun's energy to power. New systems, however, are much more sophisticated and efficient, and the technology continues to improve. Although many solar panels available to individuals still only have an efficiency rating of 14 to 16 percent, the most efficient ones available in 2018 had ratings as high as 22.5 percent. In 2017, scientists created a solar panel that is 44.5 percent efficient, but it has not yet been reproduced for people to buy. The downside of higher efficiency is that the more efficient a solar panel is, the more expensive it is; however, increasing the efficiency of PV cells will result in lower operating costs for solar power systems. Because electricity purchased from power companies is relatively inexpensive, the cost of solar power must be low enough to compete before most consumers will make the transition.

THE BENEFITS OF SOLAR CELLS

"Solar cells are truly a dream power source. They are silent, highly reliable, portable, convenient, risk-free, pollution-free, recyclable, technologically elegant, suitable for utility or customer use, mass producible, long-lived, and thoroughly compatible with a sustainable economy."

–John J. Berger, author of *Charging Ahead: The Business of Renewable Energy and What It Means for America*

John J. Berger, *Charging Ahead: The Business of Renewable Energy and What It Means for America*. New York, NY: Holt, 1997, p. 53.

Home Solar Power

Homeowners have also begun to use PV cells for their electricity needs. Although there are up-front costs of installing the solar cells on the home's roof, there are also many incentives offered by the government to encourage the use of solar power. Tax credits, rebates, and grants help homeowners in some states pay off the cost of a system. For instance, in 2015, the U.S. government passed a tax incentive that will be in effect in some form until at least 2022 that allows people who buy a solar panel system to get back up to 30 percent of the cost of their solar panels. However, even with these benefits, solar systems can cost several thousand dollars for the panels and installation.

The smartflower (shown here) is a new type of solar power system that can be used by a home or business.

Another type of home solar system is called the smartflower. Unlike traditional solar panels, which are flat, the smartflower is, as the name suggests, shaped like a flower. Multiple solar panels form the petals, and like a real flower, the smartflower can move to follow the angle of the sun, so it always generates the maximum amount of energy. According to its creators, this makes it 40 percent more efficient than other solar panels. At night, it stacks all the petals on top of each other, and in the morning, it opens automatically. As it opens, special brushes on the underside of each panel remove dust, snow, and any other material from the panel underneath it. This improves its efficiency and saves homeowners from having to climb up on their roofs to brush off the panels, as they would have to do with traditional ones. A professional must install it, but after it is installed, it works automatically.

The benefits of a home or business solar power system are considerable. Not only does the home consume significantly less energy, the homeowner or business owner can actually sell excess power generated to the local utility company. Modern home PV systems are connected to the power grid that supplies electricity from a power plant. In times of extended cloudy weather or large power needs, the home can still receive power through the grid. However, when producing more power than the home consumes, the flow of electricity reverses, sending power back to the utility company. Federal law dictates that all U.S. power companies must pay the homeowner or business owner for the power generated, but many have complained that the amount they receive for their energy is much smaller than the amount they would pay for it. For example, in Marin County, California, according to ABC News, homeowners are paid only about 4 cents per kilowatt hour (kWh). The local power company said this is because the California Public Utilities Commission sets the rates and homeowners are paid the same amount as any other energy supplier would be. However, because they feel they are being taken advantage of, many Marin County homeowners

Light Up the World

The Canadian nonprofit organization Light Up the World is working to change the lives of rural villagers who have no electricity for lighting. While on a trek in Nepal, electrical engineering professor Dr. Dave Irvine-Halliday visited a villager's hut and was amazed by the lack of light and thick smoke swirling from the cooking fire inside. Thoughts of the visit nagged him as he returned home, eventually inspiring him to develop a source of simple and economical light that could be used in those remote places.

After much experimentation, Irvine-Halliday developed a system that connected simple lights called diodes to batteries charged by solar panels. When he returned to Nepal to test his system, it was so well received that he established Light Up the World to bring light to the millions of people who live in darkness. Light Up the World is working with partners to bring down the cost of the system's components and provide grant funding so even the poorest villagers can benefit. Volunteers travel to countries such as Guatemala, Peru, Costa Rica, and Papua New Guinea to help install the panels. According to the organization's website, access to energy increases access to education by allowing children to study at night after their daily chores are done; decreases living expenses by eliminating the need to spend money on expensive, inefficient lighting methods such as kerosene; and benefits the environment as well as people's health by eliminating toxic gases created by burning fossil fuels.

with solar panels have been plugging more things in to use their excess energy, which some say is a waste of energy.

Using the Wind

Wind technology has changed a great deal since the times of the classic Dutch windmills. Instead of using wide, wooden sails mounted on broad, stout bases, energy from wind is now gathered by tall, sleek machines called turbines. The turbine is topped by a rotor made up of two or three long blades that spin in the wind. The moving rotor spins a drive shaft connected to a generator that converts the energy from the wind's movement into electricity. The electricity is sent through cables to a power substation from which it is distributed to distant cities.

Wind turbines are often built in clusters called wind farms. Proper placement of a wind farm is critical. Wind does not blow evenly in every location; some places are much more consistently windy than others. Additionally, wind speed changes throughout the day and from season to season. Wind farms need to be situated far from buildings and land features that can disrupt wind flow and create turbulence that can damage a turbine. All of these factors make it difficult to determine the best site for a wind farm. Wind scientists use complex mathematical calculations to account for these factors when selecting a site. Some of the best sites for wind farms are along coasts, lakeshores, and low-elevation ridgelines. New projects have even installed wind farms out at sea where strong winds constantly blow.

Wind farms are places that have many turbines close together to capture wind energy and deliver it to nearby houses and businesses.

Wind scientists have also contributed a great deal to increasing the efficiency of generating power from the wind. Understanding the dynamics of wind allows for turbines that are better designed to catch more wind and convert it to electricity. Sophisticated computer controls constantly fine-tune the turbines, automatically adjusting the pitch of the blades to respond to changes in wind speed or direction. Turbines designed in the early 1980s generated an average power of only about 50 kilowatts (kW). The average modern turbine now, however, can produce more than 6 million kWh in a year, which is enough to power 1,500 European households, according to WindEurope (formerly the European Wind Energy Association). The average American turbine, according to National Wind Watch, can power about 330 households in the United States per year. Offshore turbines, which are installed in the ocean, can power more than 3,300 European households. Although offshore turbines generate more power, they cannot all be built in the ocean because it takes longer to develop an offshore wind farm. By building onshore turbines, people can take advantage of wind energy while offshore turbines are still being built.

Wind turbines that generate that much power are very large. Commercial turbines already in widespread use have blades about 164 feet (50 m) long, giving the rotor a diameter of about 328 feet (100 m)—nearly the length of an American football field. Some are a little bigger or smaller, depending on how much energy they produce—a unit that is measured in megawatts (MW). The average European wind turbine produces about 2.5 to 3 MW, while the average American one produces about 1.5 MW. One MW is equal to 1 million watts, and 1 kW is equal to 1,000 watts. A watt is a measure of the rate at which energy flows. According to the website BuildingGreen, "Watts are basically the miles-per-hour measurement of the electrical world—they tell you how fast the electrons are speeding down the highway."[38]

Turbines are mounted on towers that are about 328 feet (100 m) tall. Since wind blowing a few hundred feet above the ground may be up to five times stronger than that at the surface, these tall towers place the rotors in the best position for catching the most wind possible. The taller the turbine, the more energy it can capture and the more efficiently it can work. Large wind farms may consist of hundreds or even thousands of turbines covering hundreds of acres of land. These large wind farms generate enough electricity to power an entire town.

Not all turbines are so big. Farms, small businesses, and homes also use wind turbines to provide power directly to their own electrical systems or to provide power if they live off the grid. The cost to purchase and install a personal turbine is about $30,000. The average personal wind turbine for an American home has a diameter of about 18 feet (5.5 meters) and generates about 5 kWh, but this number varies depending on the wind conditions where a person lives and how much energy they consume. According to *Windpower Engineering & Development*,

> *Small turbines range in size from 20 watts to 100 kilowatts (kW). The smaller or "micro" (20- to 500-watt) turbines are used in a variety of applications such as charging batteries for recreation vehicles and sailboats …*
>
> *A typical home uses approximately 10,000 kilowatt hours (kWh) of electricity per year (about 830 kWh per month) … A 1.5 kilowatt wind turbine will meet the needs of a home requiring 300 kWh per month in a location with a 14-mile-per-hour (6.26-meters-per-second) annual average wind speed.*[39]

TRANSFORMING THE ENERGY SYSTEM

"It's a major psychological and cultural challenge for the environmental and conservation movement. What we need to combat climate change is a complete transformation of our energy system, and that requires a lot of new stuff to be built and installed, some of it in places that are relatively untouched."

–Stephen Tindale, former executive director of Greenpeace UK

Quoted in Heather Timmons, "A Renewable Source, and Clean, but Not Without Its Critics," *New York Times*, August 3, 2006, p. C1.

The Economics of Solar Power

While the use of solar power may not yet have moved into the mainstream in the United States, worldwide use of solar power is growing quickly. The global solar industry grew 95 percent in 2016 alone and has been called the fastest-growing U.S. industry. Other countries have increased their use of solar power as well. As of 2018, China is the world's leader in solar power production—something that has happened only within the last decade. This has been especially beneficial for rural farmers, who use solar panels installed in the countryside, where there is no electric grid, to run their agricultural equipment. However, even though the largest solar farm in the world is in China and the country installed more than 42 gigawatts (GW) of solar capacity in 2016 and 2017, solar power still only accounts for 1 percent of China's total energy usage. About 66 percent still comes from coal, which is why China has a lot of air pollution. The National Energy Administration has stated that its goal is to provide 86 percent of the country's energy from renewable sources, with one-third of that amount coming from solar power.

However, one problem with solar power—worldwide, not just in China—is the cost. The price of solar technology has dropped considerably since it first became available, and according to *Forbes* magazine, it will continue to fall to eventually become the cheapest way to produce power in many countries. This is partially because "the cost of installing solar panels at big solar farms and on rooftops will drop 60% to an estimated average of around four cents per kilowatt hour by 2040 … That's cheaper than coal and natural gas power in many regions."[40]

However, as of 2018, converting to solar power was still more expensive than using fossil fuels. In discussing renewable energy, a discussion of infrastructure is crucial. Infrastructure is "a term architects, engineers, and urban planners use to describe essential facilities, services, and organizational structures for communal use, most commonly by residents of cities and towns."[41] People's taxes generally pay for infrastructure systems, and although the power created by alternative energy sources is cheap, converting the existing infrastructure to be able to accommodate alternative energy is often expensive. Setting up the means to capture the energy without creating the infrastructure to distribute it is useless: A solar panel that is not connected to anything will not transmit power. In addition to the cost, finding the right site for a solar power plant can be a challenge.

The energy infrastructure that is currently in place is made to work with the steady supply of energy that comes from fossil fuels. It is easy to control because people can add or subtract coal, gas, and oil depending on how much energy is needed. However, the weather is harder to control; on cloudy days, less solar energy is generated, while on extremely sunny days, there may be too much power to use. In California, for instance, there were 14 days in March 2017 when the state generated so much solar power that it had to pay Arizona to take some of it to avoid overloading its power lines. Cutting back on excess solar power is called curtailment. The state cannot give the power away for free because it needs to be gotten rid of quickly to avoid overloading power lines and causing blackouts. When that happens, "Arizona has to curtail its own sources of electricity to take California's power when it doesn't really need it, which can cost money. So Arizona will use power from California at times like this only if it has an economic incentive—which means being paid."[42] Because fossil fuel plants need to keep functioning for those times when solar power is unavailable, solar plants are generally curtailed first because it is more difficult and expensive to start and stop a natural gas plant. This means that when other states take California's excess energy, they are curtailing their own solar production, so carbon emissions remain unaffected.

One solution to this problem is to store the excess power in batteries and release it as needed when the sun is not shining. According to *Popular Science*,

> *Batteries ... can act as a reliable backup source. In the event of a power outage, a battery can be turned on quickly to compensate ... Batteries can also store extra energy. If there's excess power during peak wind or solar production, a battery can store up all that energy for future use. In this way, a battery is essentially acting as an end user of power as well as a power plant.*

> *But right now, this type of dual functionality doesn't fit into the way current energy grid systems are set up, says [David Hart, the director of the Center for Science, Technology, and Innovation Policy at George Mason University]. You are either one [or] the other (an end user or a power plant) and a battery is both.*[43]

Although the price of batteries, like the price of solar panels, is dropping, they are still relatively expensive. Current infrastructure can be used for now, but it is inefficient and still requires fossil fuels as a backup. Changing the infrastructure to accommodate renewable sources will carry an extra cost. All of this means people often pay higher electricity bills when their power comes from renewable energy sources. Germany broke a record in 2017 by becoming the first country to get a minimum of 35 percent of its power from renewable sources, including wind, and it is hoping to get rid of all its nuclear plants by 2022. According to the news website Reuters, electricity prices in Germany—where as much as 100 percent of the power comes from renewable resources when the days are very sunny or windy—are at an all-time high: "The run-away expansion of wind turbines and solar panels has made German prices the highest in Europe since 2013, not just because of surcharges but because more volatile [unstable] green power capacity also necessitates new transmission grids and higher costs to manage them."[44] The percentages vary depending on how much sun or wind is available on any given day. This type of increase in prices has the biggest effect on people who are already poor. As the British newspaper the *Spectator* explained,

> *Britain's environmentalists proudly announce that households have reduced their electricity consumption by almost 10 [percent] since 2005. They seldom mention that this is helped by a 50 [percent] increase in electricity prices, in part to pay for Britain increasing its share of renewables from 1.8 [percent] to 4.6 [percent]. Such a price increase of course hits the poor hardest. As with many green taxes, it does so because it taxes a basic necessity that makes up a larger proportion of a small budget. Not surprisingly, higher energy prices mean the poor are forced to reduce their electricity consumption far more than the richest, who haven't reduced their electricity consumption at all.*[45]

Forbes pointed out that in most countries, the people who continue to use only traditional energy sources end up paying more than people who have solar panels on their houses, which means their electricity bills subsidize, or reduce the cost of, the reliable energy that people with solar panels enjoy when the sun is not shining. This disproportionately affects poorer people who cannot afford to install their own solar panels.

People in developing countries also suffer from many renewable energy policies. For instance, many activists around the world are calling for fossil fuel plants to be shut down and completely replaced with renewable sources. Developing countries rely mainly on traditional energy sources to fuel their economic development, but they are responsible for much fewer carbon emissions than developed countries. Shutting down their gas and coal plants would freeze their economies while having little effect on overall emissions. Some experts say developing countries will partially subsidize the Western world's switch to renewable power and that renewable technology should not become widespread until cheap, efficient infrastructure exists to support it and reduce the burden on low-income people.

THE EFFECT ON THE POOR

"Purposefully driving up the cost of energy, while millions of Americans already struggle to pay their utility bills, is irresponsible. We cannot cut the world's carbon emissions alone, but we can certainly make U.S. energy poverty a full-blown crisis if we continue on our current course."

–Michael Jensen and William Shughart, researchers in business and finance

Michael Jensen and William Shughart, "How Green Energy Hurts the Poor," *Detroit News*, August 4, 2016. www.detroitnews.com/story/opinion/2016/08/04/green-energy-hurts-poor/88282074/.

The Economics of Wind Power

Both small and large wind turbines are increasing in use all over the world. In fact, wind power is one of the world's fastest growing energy sources; wind and hydropower (power that comes from water) combined make up two-thirds of the overall increase in renewable energy. Many countries around the world continue to invest in wind power. As of 2018, China produces the most and is continuing to build turbines. The United States is second; according to the *Energy Business Review*, "As of 30 June 2017, the country had over 52,000 commercial wind turbines operating in 41 states plus Guam and Puerto Rico. In March 2017, Xcel Energy announced plans to build 11 new wind farms across seven states in the US, with a total capacity of 3.38 GW [gigawatts]."[46] European countries are also investing heavily in wind power.

One reason for the increased use of wind power is the falling cost of capturing the wind's energy. The cost of renewable energy in general has decreased to a level where it is expected to be competitive with gas- and coal-powered electrical plants by 2020. According to a 2018 report by the International Renewable Energy Agency (IRENA), "the cost of generating power from onshore wind has fallen by around 23% since 2010 while the cost of solar photovoltaic (PV) electricity has fallen by 73% in that time."[47]

There are also significant environmental and social benefits to generating electricity from the wind instead of traditional power plants. Wind power creates no waste, pollution, or greenhouse gases. According to the American Wind Energy Association (AWEA), in 2016, wind power prevented about 159 million metric tons of carbon dioxide from being released into the atmosphere, which is equivalent to the emissions of 33.7 million cars. Unlike power plants that need water for cooling, wind farms have no need for water and can therefore be built in arid, or dry, regions. Smaller wind turbines are cost-effective for bringing power to remote villages in the world's developing regions.

However, there are some problems with wind power as well. Most obviously, wind does not blow all the time. This can be addressed when traditional power plants or solar power join wind power to form an integrated, hybrid energy system. Additionally, like solar power systems, some wind systems can store excess energy in a battery to be used when needed.

Home wind turbines can provide an extra source of clean energy, but they are generally only cost-effective if someone lives in an extremely windy place.

Another obvious issue is the cost. As with solar systems, the upfront cost is still high, so installing a personal wind turbine is still not an option for most people. Additionally, in several government tests, many small wind turbines designed for homes performed poorly; some even used more energy than they provided. However, since this is a relatively new economic field, there is no real standardization for wind turbine performance ratings, so some companies can get away with overestimating how well their product will work. According to Solacity Inc., a maker of wind turbines, personal turbines generally only make sense if someone lives in a very windy place and they pay a lot for traditional electricity, especially if their local government provides rebates, or discounts, that make installing a turbine cost-effective.

This is tied to the issue of efficiency. In a perfect world, wind turbines would convert 100 percent of the wind around them into energy, but due to certain physical limitations, the average wind turbine only converts about 50 percent of wind to energy even when operating at peak efficiency, and more frequently, their efficiency is closer to 35 to 45 percent. In comparison, coal plants are about 85 percent efficient at full capacity. This means there needs to be a lot of wind turbines to produce the same amount of energy as one coal plant.

The Politics of Solar Power

Politics also plays a large role in the adoption of renewable energy, especially in the United States, and it is often closely linked to economics. For example, one of the reasons fossil fuel energy is currently affordable for consumers is that the U.S. government gives subsidies and tax breaks to large companies to help them pay for drilling, mining, and transportation. If the government were to switch to subsidizing clean energy, the cost for fossil fuels would increase and the cost for renewable sources would decrease, allowing more people to afford them. However, there are many complex political reasons why this would be difficult, including oil and gas companies that donate money to politicians' campaigns to encourage them to vote to subsidize fossil fuels. The fact that many renewable energy companies are small and have fewer resources available limits their ability to compete with larger oil and gas companies. Not only are they less able to successfully sway politicians, they have to "compete with wealthier industries that benefit from existing infrastructure, expertise, and policy. It's a difficult market to enter … Increased government investment in clean energy—in the form of subsidies, loan assistance, and research and development—would help."[48]

Fossil fuel companies have also spent decades knowingly spreading false information about the effect their products have on climate change. With concern for the environment taken out of the picture, the slow, expensive transition to alternative energy looks much less appealing to the average consumer. Many members of Congress deny that climate change exists thanks

The California Legislature has had to fight with regulators within the state about energy sources, which has slowed down alternative energy progress.

to these campaigns, so they are less likely to support policies that benefit renewable energy. According to the Union of Concerned Scientists, one way to even the playing field for renewable energy sources would be to impose fees on companies that put out a certain amount of carbon emissions, which would make renewable energy sources more attractive to businesses.

Politics are also a reason for the difficulty of creating better infrastructure. For instance, although California produces a lot of energy from renewable sources, it cannot use all of it, mainly because of infrastructure limitations. These limitations are difficult to address because of conflicts in California's government. While the California Legislature has mandated that half of the state's power come from renewable sources by 2030, the state's regulators, who are not part of the Legislature, have been approving the construction of new natural gas power plants—a move that critics say will only benefit utility companies—instead of putting money toward improving renewable energy infrastructure. According to Jaleh Firooz, a power industry consultant, the executives who run these natural gas companies exaggerate the unreliability of alternative energy sources: "In their assumptions, the amount of capacity they give to the solar is way, way undercut because they have to say, 'What if it's cloudy?' … That's how the game is played. You build these scenarios so that it basically justifies what you want."[49] The Union of Concerned Scientists agrees that solar and wind power are much more reliable than opponents say they are, especially when combined with battery technology and smart appliances.

Concern over the availability of jobs is yet another factor that is tied to politics. This issue in particular played a large role in the 2016 U.S. presidential election, when Democratic candidate Hillary Clinton promised to close down coal companies in favor of clean energy. This statement lost her support in states where coal makes up a large part of the economy. Fear of unemployment causes many people to reject the transition to clean energy, but the reality is that clean energy will create more jobs than it destroys. In January 2017, *Forbes* reported that the solar industry alone employs more people in the United States than the oil, gas, and coal industries combined, and if renewable energy industries keep growing, the number of jobs will increase. Some residents of coal country are aware that it will be difficult for coal to make a huge comeback and boost their economy, so they are looking for other ways to make a living. This is a difficult decision for many in areas where coal mining is a longstanding family tradition; as journalist Courtney Balestier wrote, "It's hard to overstate—and perhaps, to outsiders, hard to explain at all—the mental shift that this economic change represents, and the reevaluation of identity it prompts."[50]

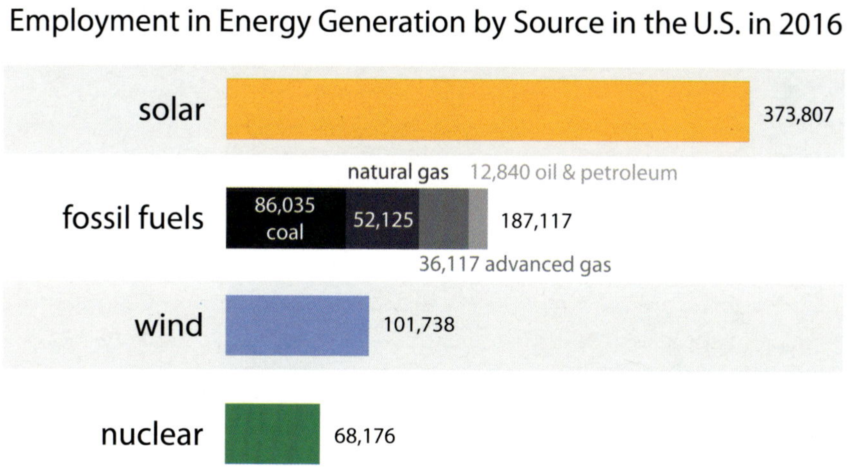

The solar industry currently employs more people than all forms of fossil fuels combined, as this information from Forbes *shows.*

The Politics of Wind Power

As with solar power, some people oppose wind power for various reasons—although often not for environmental reasons. One major complaint is that turbines are ugly. Many people who support the idea of wind power do not actually want to live near a wind farm or own their own turbine.

Wind farms are very visible and often take up a lot of space. Therefore, they dramatically change the landscape where they are built. Because areas with good wind are also often quite scenic, the installation of wind farms sometimes creates serious opposition from people in surrounding communities. Fortunately, if

A Question of Ethics

Researcher and environmentalist Peter Harper studies the environmental impacts of both fossil fuels and renewable energy alternatives. Harper agrees with critics of wind farms who say they contribute to a decline in the beauty of the landscape. However, he maintains that in sharp contrast to the use of fossil fuels, clean, safe, renewable wind energy should be embraced for ethical reasons. In an article entitled "Why I Hate Wind Farms and Think There Should Be More of Them," he wrote,

> *Nearly all objectors to wind-farms live in rural areas and don't want their landscapes spoiled. This is quite understandable; neither do I. But such people invariably wish to retain the benefits of abundant cheap energy, brought to them discreetly by means of cables, pipelines and tankers, for whom others (mostly in urban areas or overseas) have borne the impact of mining, refining, conversion, transmission etc, not to mention the wider implications of CO2 release. This is simply hypocritical. In opposing windpower on grounds of personal aesthetics, objectors are implicitly supporting something far worse, and trying to avoid paying their dues.*[1]

1. Peter Harper, "Why I Hate Wind Farms and Think There Should Be More of Them," *ReNew*, no. 107, May/June 1997

wind projects can overcome this opposition, they seem to stand a good chance of success, according to the *Des Moines Register*:

> *Once turbines go up … people generally get over their initial reservations, said Lu Nelson, a policy program associate at the Center for Rural Affairs in Lyons, Neb.*
>
> *"A lot of it is the fact that these are new for people, even if you live in Iowa," he said. "After some time, it's just part of the landscape, and it's not this intrusive thing anymore."*[51]

To address some of the aesthetic, or beauty, concerns, some companies are making turbines that look like trees or art sculptures. At the Oklahoma Medical Research Foundation building, wind turbines were installed that look like strands of deoxyribonucleic acid (DNA)—inherited material that determines things such as how someone looks and what diseases they may have. These turbines help power the building, look interesting, and are related to the work that goes on in the building. However, some people believe people should worry less about the beauty of turbines and simply be happy with the clean energy. "There's no free lunch. 'Not in my backyard' is not environmentalism,"[52] said Paul Hansen, former executive director of the Izaak Walton League of America, a conservation organization.

Researchers have suggested that painting wind turbines a color other than white could reduce the number of wildlife that die in the blades.

However, one proposed change in the way turbines look may actually eliminate another issue with them. Despite the fact that wind power does not contribute to the environmental damage associated with the use of fossil fuels, it does have a different negative effect on the environment: Wildlife such as birds and bats are sometimes killed by the turbines' spinning blades. One study determined that this is likely because most turbines are white to make them more visible to airplanes while at the same time helping them blend in with the landscape. However, the light color attracts insects, which in turn attract wildlife that eat those insects. It has been proposed several times that turbines should be painted a different color, such as purple, which experts say would make them less attractive to insects and wildlife while making them more interesting for people to look at—although some people have said they would view turbines as uglier if they were purple, and they do not think the number of wildlife being killed each year is high enough to justify this change. The researchers in the color study also noted that other factors, such as the heat turbines generate and potential interference with bats' echolocation, may also be contributing to the problem, so changing the color may not be a perfect solution.

A third, relatively new objection to wind turbines is the belief that they are causing health problems called "wind-turbine syndrome" in people who live near them. A wind turbine's blades can spin up to 180 miles (289.7 km) per hour. According to some experts, "In high winds, this rapid spinning can produce sound and vibration—in addition to disruptions in air pressure."[53] People who are concerned about wind-turbine syndrome believe the spinning causes infrasound, which, according to *The Atlantic*, is sound that is believed to be such a low frequency that it is inaudible to humans, although if the volume gets high enough, people can hear it. When the volume is high enough for people to hear, it can indeed have dangerous health effects, simply because the volume has to be so incredibly high in order for people to hear it. However, what people are concerned about in the case of wind turbines is that noise they cannot hear may be causing their symptoms.

The symptoms of wind-turbine syndrome—which is not a recognized medical illness—are said to include headaches, irritability, anxiety, night terrors or learning disabilities in children, ringing in the ears, and problems with concentration and memory. Although some studies have been done, they have not determined for sure whether wind-turbine syndrome exists. It has been proven that wind turbines do produce an audible noise and vibration at high speeds, but there are mixed opinions on whether they produce more noise than things such as diesel engines or waves crashing on the beach, neither of which are considered infrasound or produce the symptoms supposedly associated with wind turbines. There have been some documented reports of people who live near turbines having difficulty sleeping, but chronic sleep loss is known to affect some people who live near any source of noise, such as a highway. Long-term loss of sleep can also cause the other symptoms associated with wind-turbine syndrome.

Some psychological studies have also suggested the symptoms may be caused simply by the power of suggestion. *Slate* magazine described the results of one such study:

> *Researchers exposed 60 participants to 10 minutes of infrasound ... and sham [fake] infrasound (that is, silence). Before the listening sessions, half the group was shown television footage of people who lived near wind farms recounting the harmful effects they said were caused by noise from the spinning blades. Within this group, the people who scored high on a test of anxiety became symptomatic whether they were exposed to low-frequency noise or sham infrasound.*[54]

Experts have noted that symptoms of wind-turbine syndrome were almost nonexistent until it began to be mentioned in the media. The more it was mentioned, the more complaints and symptoms were recorded, leading many to believe the condition is not real. However, to address people's claims of illness and the known noise problem, current wind technology uses sound-dampening systems, and space buffers of up to 2 miles (3.2 km) are recommended between wind farms and residential areas. Experts hope these measures will be enough to resolve the issue so people can keep experiencing the energy benefits of wind farms.

Using the Earth's Resources

Solar and wind power are not the only forms of alternative energy being harnessed to create electricity. The power of the oceans, heat from the earth, and radioactive material are all sources of energy that are already being used, although some of these technologies are more developed than others. Each has its benefits and drawbacks as well as its opponents and supporters.

Hydropower

Hydropower is energy that comes from harnessing the movement of water. Anyone watching the rushing of a river or the crashing of waves onto the seashore has witnessed the power that moving water can create. People have been using hydropower since the beginning of recorded history, often by damming or diverting the flow of rivers. The water wheel was also invented to harness waterpower. In modern times, the flow of water is directed through rotating turbines to produce clean and renewable electricity. According to the Institute for Energy Research, in 2015, renewable energy accounted for about 13 percent of America's overall power; hydroelectric power from dams and large waterfalls,

Water wheels have been used for thousands of years to provide energy.

such as Niagara Falls, provided about 46 percent of that total. Worldwide, hydropower accounts for about 17 percent of all energy production. Hydropower has many benefits: it is a clean source of energy, it can be used in any state with enough water (which decreases reliance on foreign energy sources), the lakes created by reservoirs and dams are often open to the public for boating and swimming, and they provide flood control and irrigation for crops.

Percent of Total Energy Produced by U.S. Power Plants

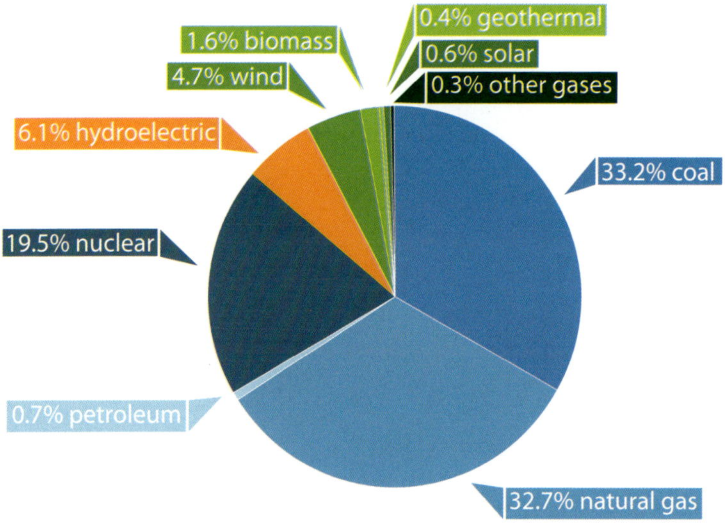

Although hydropower is a small percentage of overall energy production in the United States, it is a large percentage of currently used renewable energy sources, as this information from the Institute for Energy Research shows.

Power created by diverting rivers and building dams, however, often comes at a significant cost to the environment. When the flow of a river is altered, the entire ecosystem is affected. It becomes difficult or impossible for fish and other wildlife to travel past dams that block historic migration routes. The habitat downstream from a dam changes drastically with decreased or irregular water flows, and building a power station can change the temperature of the water, which can also have a negative effect on fish and water plants. The creation of a dam floods entire valleys, which displaces wildlife and destroys habitats.

Other Sources of Hydropower

Because of their associated environmental concerns, traditional hydroelectric power projects have been decreasing in acceptance, and many places are choosing not to build more. For instance, in 2016, Cambodia's minister of mines and energy declared that no new dams would be built until at least 2020 due to their high upfront cost and negative effects on both wildlife and local villages. Although some dams are still being built in some places, more governments are choosing to invest in solar and wind power. Additionally, more innovative ways to harness the power of water are being explored. Instead of looking to rivers for power, scientists are now experimenting with using the tremendous force created by the movement of ocean waves and tides to produce electricity.

Scientists are testing multiple strategies for capturing energy from ocean waves, although some of them are still experimental and can only currently be found in one or two places. According to a 2014 article in *Yale Environment 360*, a journal published at the Yale School of Forestry & Environmental Studies,

> *Wind and solar power have taken off in the past decade or two … Meanwhile, numerous studies have concluded that wave power—and to a lesser extent, tidal power—could contribute massive amounts to the overall energy picture. But while the industry has made halting progress, experts agree that it remains decades behind other forms of renewables, with large amounts of money and research required for it to even begin to catch up.*[55]

Some small-scale wave power installations have been tested, but a major challenge facing wave power is the difficulty of designing a system that will efficiently capture the power. Many different designs have been tried, including buoys that bob on the surface of the water as well as devices that are attached to the ocean floor. One design is a 450-foot-long (137 m) snake-like steel tube made up of hinged segments. The train-sized "sea snake" is anchored to the ocean floor at both ends. As the jointed segments move back and forth in the waves, the motion pushes pistons that drive electricity generators. The electricity is sent to shore through underwater cables. Other devices "pass the mechanical energy of the wave along to land before turning it into electrical energy."[56] According to *Yale Environment 360*, it is still unclear which device is the best, and more research will need to be done. However, experts in the wave power industry believe that if steady progress is made, wave power can be connected to the grid by 2035.

Another way to get power from the sea is to take advantage of the large volume of moving water during tidal flows. There are several different ways to get power from the tides. One is called a barrage; it is a structure similar to a dam that is installed in a place with strong tides, such as the inlet of an ocean bay. The U.S. Energy Information Administration explained, "Sluice gates on the barrage control water levels and flow rates to allow the tidal basin to fill on the incoming high tides and to empty through an electricity turbine system on the outgoing ebb tide. A two-way tidal power system generates electricity from both the

Pictured here is a "sea snake" wave power system.

Tidal turbines such as the ones shown here generate energy from the tides that flow around them.

incoming and outgoing tides."[57] Another option is tidal turbines, which look similar to wind turbines but are placed on the ocean floor, where the tidal flow is strong. Tidal turbines cost more than wind turbines because they have to be sturdier, but they generate more energy.

The strength of tides varies a great deal around the world. Strongest tidal flows generally occur where narrow gaps in land formations funnel large amounts of water through a small area. For instance, some of the fastest tides in the world are in Skookumchuck Narrows in British Columbia. As of 2018, tidal turbines are in place in several locations, including Scotland and South Korea, and another project is being developed for the East River in New York.

Engineers involved in testing wave and tidal generators find considerable promise in the technology. Waves and tides are much more consistent than wind and hold significantly more power in a concentrated area. Wave machines also leave a much smaller mark on the landscape. They are quiet, do not block views, and require much less space than wind farms. Although they could pose a hazard to ships if not clearly marked or placed outside of shipping lanes, they pose little threat to marine wildlife. Tidal machines, too, have been found not to pose a long-term disturbance to sea creatures and birds that live near water.

Heat from the Ground

An alternative energy industry that is already well established is geothermal power. Geothermal power uses heat generated from deep within the earth's core. Geothermal power is most easily accessible in parts of the world with volcanic activity. In these places, heat from the earth's core rises through fissures, or deep cracks in the earth's surface, to reach the surface, where it heats rocks and water.

The resulting hot water is used to power geothermal plants. Geothermal plants generate electricity by using steam from hot water to drive turbines. The world's largest geothermal power plant is near San Francisco, California. It is called the Geysers Complex, and it is actually a network of 22 plants that work together to produce electricity for California's Sonoma, Mendocino, and Lake counties. The Geysers Complex has produced electricity commercially since 1960. It gets its steam from more than 350 wells drilled as far as 2.5 miles (4 km) into the rock. Water and steam emerge from the wells at temperatures between 250 and 700 degrees Fahrenheit (120 and 370 degrees Celsius).

Unknown Potential

Author John J. Berger has studied the potential for renewable energy growth in the United States. He believes that if technology advances, one day, environmentally friendly geothermal power could supply all the energy America would ever need. For now, however, this energy potential is largely untapped, mostly due to a lack of research funding and the availability of cheaper fossil fuels. Berger acknowledged that no one knows how much power is available from geothermal sources. He said, "By one estimate, geothermal resources exceed the world's entire coal, oil, and natural gas resources by 30 times. Another estimate found the heat content of the Earth's crust is hundreds of thousands of times U.S. annual energy demand."[1]

1. John J. Berger, *Charging Ahead: The Business of Renewable Energy and What It Means for America*. New York, NY: Holt, 1997, pp. 234–235.

There are three different types of geothermal plants. As *National Geographic* explained,

> *Dry steam, the oldest geothermal technology, takes steam out of fractures in the ground and uses it to directly drive a turbine. Flash plants pull deep, high-pressure hot water into cooler, low-pressure water. The steam that results from this process is used to drive the turbine. In binary plants, the hot water is passed by a secondary fluid with a much lower boiling point than water. This causes the secondary fluid to turn to vapor, which then drives a turbine. Most geothermal power plants in the future will be binary plants.*[58]

Geothermal energy, like wind and solar power, is a clean form of energy that does not rely on fossil fuels. However, unlike weather-based energy, it is always available. It is also much cheaper to produce than energy that comes from fossil fuels, especially when used as a direct heat source. This is when hot water from the ground is pumped through pipes, and it can have various uses. Some communities already use geothermal heat extensively; for instance, some American cities pipe water under roads to melt ice and snow in the winter. In Iceland, which has at least 25 active volcanoes as well as hot springs and geysers, many buildings and swimming pools are heated with geothermal energy, which makes it one of the world's cleanest countries.

Geothermal energy plants are often built near active volcanoes because the heat from the earth's core is closer to the surface in those spots.

As with all energy sources, however, geothermal energy has its drawbacks:

> *The main concern is the release of hydrogen sulfide, a gas that smells like rotten egg at low concentrations. Another concern is the disposal of some geothermal fluids, which may contain low levels of toxic materials. Although geothermal sites are capable of providing heat for many decades, eventually specific locations may cool down.*[59]

Nuclear Energy

The quest to develop new options for replacing fossil fuels has led to much debate between supporters of the various alternative energy sources. No source, however, is more intensely debated than nuclear power. Some people consider nuclear power to be a clean, safe, and affordable way to meet the world's growing energy needs. Others strongly object to the use of nuclear energy due to concerns about power plant safety, the storage of dangerous nuclear wastes, and the potential for nuclear materials to be used as weapons.

Nuclear energy is created when the nucleus at the center of an atom is split in a process called nuclear fission. Nuclear fission uses the element uranium as fuel. Uranium is a fairly common element on Earth, originating from the remains of shattered stars present when the planet was formed. However, the type of uranium that is used for nuclear energy, which is type U-235, is extremely rare—less than 1 percent of all uranium is type U-235. The nuclei of U-235 are unstable and split easily. When a uranium nucleus is split, a great deal of energy is released in the form of heat.

Nuclear reactors work by controlling the fission, or splitting, of uranium atoms. The uranium fuel is formed into small pellets that are arranged into long rods inside the reactor. The uranium fuel rods are inserted into a water-filled chamber called the reactor core. Once inside the core, the uranium is bombarded with particles called neutrons. When a neutron collides with a uranium nucleus, the nucleus breaks up, releasing more neutrons. When some of these neutrons hit another uranium nucleus, that atom will also break apart and release even more neutrons. Some of these neutrons strike still more uranium atoms, splitting those as well. This process is called a chain reaction.

If a chain reaction continued uncontrolled, the uranium would create so much heat that it would melt. To prevent this, the fuel rods are separated by control rods of a neutron-absorbing material, which is called a nuclear poison. Reactor operators raise or lower the amount of control rods to manage the rate of the chain reaction. To create more heat, the control rods are raised, allowing more uranium to be bombarded by neutrons. To diminish the heat, the control rods are lowered between the fuel rods to absorb neutrons and slow the reaction. The water in the core also cools the uranium and assists in slowing the reaction.

The heat produced by the chain reaction boils the water in the reactor, turning it into steam. Just as in conventional coal-fired power plants as well as geothermal plants, the steam runs through a turbine, causing it to spin a generator that produces electricity. The steam is then cooled in the nuclear plant's large cooling towers before being recirculated back to the core for reheating.

Controversy over Nuclear Power

Although nuclear power is a clean and efficient energy source when it is working properly, even one accident can be devastating to communities for miles around. Nuclear power plants have been around since the 1950s, and since that time, several tragic accidents have occurred. The most infamous occurred on April 26, 1986, at the Chernobyl nuclear power plant in the former Soviet Union (now Ukraine). Without warning, the reactor core exploded, blowing the top off the reactor and igniting a raging fire. With no special training on dealing with a reactor fire, firefighters poured water onto the blazing core. When the water hit the melting uranium, it billowed back up into the sky as radioactive steam. This radioactive fallout was carried by the wind thousands of miles over Asia and Europe, where it contaminated soil and agricultural crops.

The towns around the Chernobyl power plant had to be abandoned in a hurry and are still not safe to live in.

Closer to the disaster site, the effects were even more catastrophic. Although the Soviet Union reported 31 deaths at the time of the accident, Ukraine officials now say more than 4,300 people were killed. Additionally, radioactive exposure may be linked to more than 6,000 cases of thyroid cancer. As many as 375,000 people had to leave their homes because the area was so contaminated, leaving ghost towns behind. In addition, exposure to harmful doses of radiation may have affected as many as 17.5 million people in the region. The effects of nonlethal doses of radiation may take decades to become apparent in the form of health issues.

Accidents such as this one make many people oppose the use of nuclear power. They say it is not worth the risk. Supporters, however, are pushing for giving the technology a fresh look. In fact, even some loyal environmentalists, long opposed to nuclear power, are reconsidering their position in light of nuclear power's potential for reducing the world's dependence on fossil fuels. This surprising shift in attitude is driven mostly by a desire to halt the serious consequences of climate change caused by the burning of coal and petroleum products. Patrick Moore, a longtime environmentalist, explained his new take on nuclear power:

> *In the early 1970s when I helped found Greenpeace, I believed that nuclear energy was synonymous with nuclear holocaust, as did most of my compatriots … Thirty years on, my views have changed, and the rest of the environmental movement needs to*

update its views, too, because nuclear energy may just be the energy source that can save our planet from another possible disaster: catastrophic climate change.[60]

While disagreement persists about the dangers of nuclear power, both sides recognize the troubling challenge of handling and storing nuclear waste. Hazardous radioactive waste material is produced at all stages of the nuclear process, from mining the uranium to the operation of a nuclear power plant. These wastes must be stored in such a manner that the radiation cannot escape and harm people and the environment.

The most dangerous, or high-level, waste is the uranium fuel itself. Fuel rods can power fission reactions for 1 to 3 years, after which time they are removed from the core. Some of the uranium can be reprocessed into new fuel, although this process is not yet permitted in the United States due to safety concerns. This is the only way to get more uranium; although it is a clean source of energy, it is not a naturally renewable resource, which gives it one of the same limitations as fossil fuels: Eventually, the earth will run out. Spent fuel rods are extremely hot from the fission reactions and must be stored in water until they cool. These fuel rods will continue to give off radiation for tens of thousands of years. So far, no one has discovered a safe way to store this waste that will guarantee it stays secure for such a long time period.

Nuclear power plants are highly controversial. Some people see them as a good source of clean energy, but others believe they are not worth the health risk.

As of 2018, nuclear power provides more than 25 percent of the European Union's (EU's) total electricity and nearly 20 percent of the United States's. Some governments and activists have expressed a desire to move away from nuclear power; for instance, Belgium has pledged to phase out nuclear power by 2025. However, with such heavy reliance on nuclear power, it is questionable whether this can be done in that time frame. Some question whether it can be done at all, but others believe that with the renewable options currently available, it is a feasible goal. Some even say it is an unavoidable occurrence. Antinuclear activist Stephane Lhomme said,

> *Out of the 400 reactors in service globally, give or take, three quarters of them are too old and are heading towards the end of their life spans …*
>
> *So in any case, whatever your thoughts on the matter, whether you're for, against, whether there's an alternative or not, in any case the majority of these reactors are going to close or are in the process of closing down.*[61]

If no new reactors are built to replace these old ones, the world's reliance on nuclear energy will naturally decrease. However, not all places are following Belgium's lead. In New Jersey, for instance, lawmakers passed a bill in April 2018 to preserve its nuclear industry and the jobs that industry provides. The bill will grant subsidies, or government money, to help make nuclear power plants more competitive with fossil fuels as well as wind and solar power. The debate about this issue will likely continue far into the future.

CHAPTER 5

What Happens Next?

Many issues in the energy debate are uncertain, but one thing is sure: Alternative energy is here to stay. It is already being used in many places, and researchers and entrepreneurs are spending a lot of time and money to come up with better processes—some of which refine old techniques, while others are experimental and may or may not become widespread. A 2017 study by the Pew Research Center found that 65 percent of Americans support research into developing alternative energy sources, compared with only 27 percent who support expanding the use of fossil fuels.

However, it is also clear that fossil fuels are unlikely to have a single, dominant replacement. Most energy experts agree that there is no magic solution just around the corner that will meet the world's energy needs. Instead, they predict a new global energy system in which diverse fuel sources work together to provide clean, abundant energy. In addition, this new system will have an increasing reliance on energy efficiency and conservation.

Going Green

Support for this type of hybrid system is already gathering momentum. Power generated through the use of environmentally friendly technology such as solar and wind power is sold under the label of "green power." Services such as CleanRev have been created to make it easier for consumers to access clean energy. When people sign up with CleanRev, they still get their energy from their local utility company, but it comes only from renewable sources. Traditionally, power companies purchase energy from whichever source is cheapest; this can include a mix of

Large Divide between Republicans and Democrats over Energy Priorities

% of U.S. adults who say alternative sources should be the more important priority for addressing America's energy supply

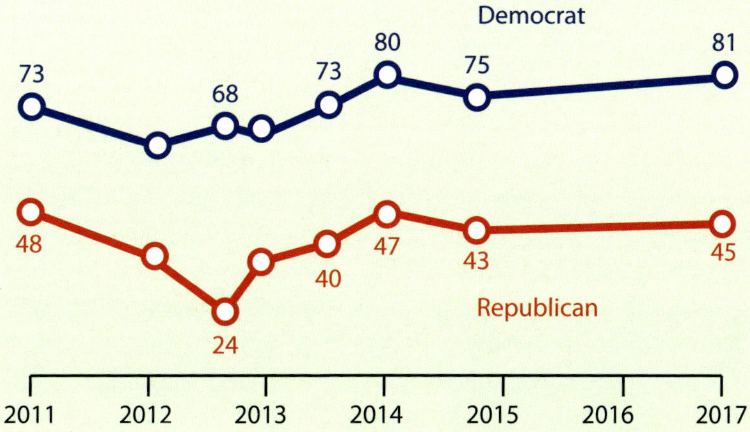

The Pew Research Center found that people's political affiliation affected their opinions about alternative energy versus fossil fuels.

renewable and fossil fuel sources. CleanRev offsets the cost of renewable energy for the company so it can provide that source alone to its customers. The more people use it, the more companies will see there is a demand for renewable energy, so they are more likely to provide it for their customers.

Many people appreciate this service because they want an easy way to use renewable energy, without installing new equipment or seeing a dramatic jump in their energy bills. Heather Kohn, who uses CleanRev to power her home in Columbus, Ohio, said it was very easy to make the switch and she has seen no real difference in the amount of her energy bill, although the way it is calculated has changed: "My rate is flat with CleanRev, instead of variable depending on what my provider, AEP Ohio, chose to charge. It makes my bills predictable based on my family's energy consumption."[62] An additional benefit, she said, was knowing that she was doing something within her means to help the environment:

> *I feel good about switching. It's an easy step to put my vote of confidence in renewable energy without the expensive costs associated*

with other options—like installing solar panels. I may choose to do something more in the future, but this is a quick way to start sending a message to the energy industry and elected officials.[63]

Making Green Cities

Encouraging individuals and businesses to turn to renewable energy sources is just the beginning of what some advocates hope will become a green energy revolution. However, many people have pledged to fight climate change without taking real action. For instance, in 2007, more than 1,000 mayors, including New York mayor Michael Bloomberg, signed a document called the U.S. Conference of Mayors Climate Protection Agreement. They disagreed with President George W. Bush's rejection of an international climate agreement called the Kyoto Protocol. Since then, the Kyoto Protocol has been replaced with what is often called the Paris Agreement or Paris accord, which President Barack Obama signed but President Donald Trump rejected. The Conference of Mayors condemned Trump's actions just as they had condemned Bush's.

However, most of the mayors in the organization have been slow to make real change in their cities, prompting some to criticize their statements as empty promises. In 2017, the *National Review* wrote that after all the cities whose mayors had signed the agreement failed to meet the Kyoto Protocol's goals, they "are replacing those old promises with new ones. Cities are now

A LACK OF UNDERSTANDING

"There's an immense and, at times, willful ignorance in America about what goes on behind the light switch. Nowadays, people can tell you exactly what they paid for their last gallon of gas, but they're clueless when it comes to electricity. Most people don't know what a kilowatt is, let alone what it costs."

–Jeff Goodell, environmental author and journalist

Quoted in Nelson Harvey, "The Real Black Gold," *American Prospect*, June 13, 2006. prospect.org/article/real-black-gold.

signing the 'We Are Still In' pledge to meet the Paris Accord's CO2 emissions targets. Their history of failure demonstrates how hollow the new promises are."[64]

One big step toward achieving the program's goals would be reducing citizens' dependence on cars by developing better public transportation systems or making cities more walkable. However, this can be difficult and expensive because it often requires changing the local infrastructure. For instance, biking is a good green alternative to cars and buses, especially in places that are warm year-round. However, many cities and towns are not safe for bicyclists, so they must be changed—for instance, by creating designated bike lanes on the road, connecting more places with bike paths, and installing racks for people to park and lock their bikes.

Changing the way a city is already laid out is generally more challenging than starting from scratch, which makes some people resist the changes. However, many places have made it work. The best example is Copenhagen, Denmark, which was named the European Green Capital in 2014 for doing things such as expanding bicycle access and adopting electric buses. The city did this in response to worsening climate change—most notably, a flood in 2011 that did millions of dollars' worth of damage. Copenhagen officials responded by creating the Copenhagen Climate Adaptation plan and setting the goal of becoming carbon neutral by 2025.

Another way cities can go green is by using architecture in innovative ways. For instance, the architecture firm Bjarke Ingels Group (BIG) often uses

Copenhagen is one city that has committed to making its infrastructure more bicycle-friendly. Because of this, many more people use bicycles there than in most other cities.

plants in its designs. This can help make the buildings more efficient—for instance, putting plants on a roof can naturally keep a building cooler in summer because it does not have a dark roof to absorb heat, so less air conditioning needs to be used, thus saving power. Cities can also increase the number of plants that are available to absorb carbon dioxide from the air. They are interesting to look at, too, which is an important factor for many people.

In 2017, China announced its plans to build a "Forest City" as a way to fight pollution. All the buildings in the city were designed to be covered with plants and trees that can absorb the carbon dioxide in the air and release oxygen. The architects plan to have construction finished by 2020.

Cities can make significant headway into reducing their production of climate-changing pollutants by using energy alternatives and increasing the use of public transportation. However, another way to reduce the need for fossil fuels is through energy conservation. Energy can be conserved in two ways. One way is when people change their behavior in order to use less energy. Using sweaters and blankets in the winter, setting thermostats slightly lower, and using fans and opening windows for summer cooling can help address this problem. Insulating homes can also help reduce both heating and cooling costs.

Compact fluorescent bulbs (middle) are more energy efficient than incandescent ones (left), but LED bulbs (right) are the most efficient of the three.

Another simple way to reduce home emissions is reducing electricity usage by buying more efficient appliances and using more efficient lighting options. Consumers can look for the blue Energy Star labels, which signify energy efficiency, and substitute compact fluorescent bulbs for incandescent ones. In fact, according to the U.S. Environmental Protection Agency (EPA), if every American household replaced one of their most-used light bulbs with a compact fluorescent light bulb, it would be the equivalent of removing 800,000 cars and would save enough energy each year to light 3 million homes. Even better, homeowners can use light-emitting diode (LED) light bulbs, which last 50,000 hours, or around 11 to 17 years, but consume only about 10 percent of the electricity used by incandescent bulbs. Home electric usage can also be reduced by simply unplugging computers, televisions, and appliances when they are not in use. Even if they are turned off, experts say, many electronic devices continue to use energy as long as they are plugged into an electric outlet.

THE ENERGY EFFICIENCY PARADOX

"[A]s a general rule it's important to understand that for all its other benefits energy efficiency doesn't necessarily mean energy conservation ... the sad truth is that no matter how efficient we make our machines, reducing our energy use is almost certainly going to mean foregoing [giving up] some of the things we could be using energy on."

–Robert de Neufville, journalist

Robert de Neufville, "The Energy Efficiency Paradox," *Big Think*, accessed on April 12, 2018. bigthink.com/politeia/the-energy-efficiency-paradox.

A Change in Attitude

While many people claim to be concerned about energy conservation, in practice, they often continue to waste a lot of energy. Many energy industry observers believe a shake-up in American attitudes about energy is needed to truly make a difference. Michael T. Klare, author of *Blood and Oil*, explained,

> *We are not likely to change our fundamental behavior—to start relying less on conventional automobiles and other oil-powered devices—unless there is a dramatic change in national attitudes*

> *about individual and societal energy behavior. We need, in short, to undergo what sociologists call a* paradigm shift*—a complete rethinking of our basic outlook on this critical issue.*[65]

Klare described how people have experienced such attitude shifts in the past; today, for example, the once-acceptable practice of smoking is banned in most public places.

A change in attitude that leads to conservation action will only be possible if it is supported by changes in the other factors currently contributing to America's oil addiction. These changes include increasing public education about the consequences of continued dependence on oil, such as global warming and health-threatening pollution. A greater focus on the issue by state and national leaders will also play an important role in increasing acceptance of conservation measures. Most important, however, will be the availability of affordable and readily available energy alternatives.

New Initiatives

New technology and new studies are being pioneered all the time in an effort to make renewable energy more feasible. One promising new technology is a hybrid solar cell that also collects energy from rainwater. This would be a huge advancement because it would make solar panels useful even when the sun is not shining. The hybrid cell combines PV cells with a device called a triboelectric nanogenerator (TENG), which gets power from the friction that is created when two materials rub together. As raindrops roll across the TENG, it generates power. According to the website Science Alert, "While the idea of using TENGs like this isn't completely new, the challenge has been to develop a system that wasn't too complex or bulky."[66] Research is still ongoing to create the most efficient hybrid system possible.

Another new, relatively inexpensive initiative is called bioenergy with carbon capture and storage (BECCS). *Science* magazine explained how it works:

> *The idea is to cultivate fast-growing grasses and trees to suck CO2 out of the atmosphere and then burn them at power plants to generate energy. But instead of being released back into the*

Saving Energy

U.S. corporations are already beginning to demonstrate a shift in their attitudes toward energy conservation. These companies are finding that incorporating energy-conserving practices into their business plans not only develops goodwill within their communities, it also benefits the company's bottom line.

Individuals can also make changes at home to save energy. The following are all free or low-cost ways to reduce energy usage:

- drying clothes outdoors when the weather is nice or indoors on a drying rack
- turning off lights and appliances when they are not being used
- taking shorter showers
- turning off the water heater before going on vacation
- plugging electronics into a power strip and shutting off the power strip when not in use
- sealing cracks around the home to prevent warm air from escaping (this can be done with caulk or weather stripping, both of which can be purchased at a hardware store)

People can see further benefits if they are able to make larger purchases, such as installing energy-efficient windows or a smart thermostat that automatically adjusts the temperature when no one is home.

atmosphere in the exhaust, the crops' carbon would be captured and pumped underground …

In some future scenarios, BECCS would remove up to a trillion tons of CO2 from the air by the end of the century—about half of what humans have emitted since the start of the Industrial Revolution—and it would supply a third of the globe's energy needs.[67]

Researchers are still working on addressing some concerns, such as whether there is enough land available to grow all the crops needed to meet these goals. There is a lot of unused farmland in the world that is currently only growing weeds, but some experts are unsure whether it is enough. They must also consider economic issues, such as how to make growing these specialized grasses and trees profitable for farmers, who would have to compete with farmers who grow food crops. However, experts say that if they get it right, BECCS could have a huge positive impact on the world.

A third program currently being implemented is a campaign called 100%, which was launched by actors Leonardo DiCaprio and Mark Ruffalo in 2015. The campaign aims to make affordable renewable energy available to everyone. Part of 100% involved a study that was released in late 2017, which created plans for 139 countries to transition completely to renewable sources by 2050. Whether the plans will be implemented—and whether they will be successful—remains to be seen.

Looking Forward

In the United States, the federal government has been generally unsupportive of alternative energy since the election of President Donald Trump; Scott Pruitt, the head of the EPA, encouraged Trump to pull out of the Paris accord and has passed many regulations to give economic benefits to the fossil fuel industries. This has made it harder for the United States to provide alternative energy at a reasonable cost to its customers. However, every day, new technologies are being discovered that make the adoption of renewable energy easier and cheaper, and some individual states and companies are taking steps toward renewable energy, regardless of the federal government's policies. Additionally, many other countries are putting resources into developing alternative energy sources, which experts say will soon mean the United States will need to adopt them as well if it wants its economy to remain competitive with the rest of the world. No one knows what new technology may be developed in the future, but as the cost of currently available alternative energy sources keeps dropping, many hope fossil fuels will soon be history.

NOTES

Introduction: The Problem with Fossil Fuels

1. Kate Samuelson, "Renewable Energy Is Creating Jobs 12 Times Faster than the Rest of the Economy," *Fortune*, January 27, 2017. fortune.com/2017/01/27/solar-wind-renewable-jobs/.

Chapter 1: Why Is Alternative Energy Necessary?

2. Paul Roberts, *The End of Oil*. New York, NY: Houghton Mifflin, 2004, p. 26.
3. Bill Bryson, *At Home: A Short History of Private Life*. New York, NY: Broadway Books, 2003, pp. 241–242.
4. Roberts, *The End of Oil*, p. 6.
5. Rob Davies, "Global Demand for Energy Will Peak in 2030, Says World Energy Council," *The Guardian*, October 10, 2016. www.theguardian.com/business/2016/oct/10/global-demand-for-energy-will-peak-in-2030-says-world-energy-council.
6. Tyler Crowe, "We Have Way More Than 53 Years' Worth of Oil Left," The Motley Fool, August 2, 2014. www.fool.com/investing/general/2014/08/02/we-have-way-more-than-53-years-worth-of-oil-left.aspx.
7. Quoted in Roberts, *The End of Oil*, p. 56.
8. Crowe, "We Have Way More Than 53 Years' Worth of Oil Left."
9. Quoted in "Steady as She Goes," *The Economist*, April 20, 2006. www.economist.com/node/6823506.
10. Roberts, *The End of Oil*, p. 65.
11. Sarah Sax, "Keystone Pipeline Oil Spill Could Be Worse Than We Thought," Vice News, November 18, 2017. news.vice.com/en_us/article/ywnn7v/keystone-pipeline-oil-spill-could-be-worse-than-we-thought.
12. Quoted in "Visibility Effects of Air Pollution," National Park Service, accessed on February 16, 2017. www.nps.gov/subjects/nightskies/visibility.htm.

13. Al Gore, "The Climate Crisis and the Need for Leadership," *Rolling Stone*, November 17, 2005. www.rollingstone.com/politics/news/the-climate-crisis-and-the-need-for-leadership-20051117.

14. Quoted in Ross Gelbspan, "Michael Crichton's Misstated State of Fear," *National Wildlife*, April/May 2005, p. 12.

15. Matthew Yeomans, *Oil*. New York, NY: New Press, 2004, p. 121.

16. Quoted in Michael T. Klare, *Blood and Oil*. New York, NY: Metropolitan, 2004, p. 9.

17. Klare, *Blood and Oil*, p. 7.

18. Quoted in Antonia Juhasz, "Why the War in Iraq Was Fought for Big Oil," CNN, April 15, 2013. www.cnn.com/2013/03/19/opinion/iraq-war-oil-juhasz/index.html.

19. "What Is Fracking and Why Is It Controversial?," BBC, December 16, 2015. www.bbc.com/news/uk-14432401.

20. "What Is Fracking," BBC.

21. Jennifer Horton, "Why Is Offshore Drilling So Controversial?," How Stuff Works, August 11, 2008. science.howstuffworks.com/environmental/energy/offshore-drilling-controversy2.htm.

Chapter 2: Rethinking Transportation

22. Quoted in Katharine Mieszkowski, "Can Hydrogen Save Us?," *Yes!*, Fall 2004, p. 29.

23. Wes Raynal, "2017 Chevrolet Bolt EV Review: Is It the Best Electric Car on Sale?," *AutoWeek*, November 8, 2017. autoweek.com/article/car-reviews/2017-chevrolet-bolt-ev-review-it-best-electric-car-sale.

24. Joey Gardiner, "The Rise of Electric Cars Could Leave Us with a Big Battery Waste Problem," *The Guardian*, August 10, 2017. www.theguardian.com/sustainable-business/2017/aug/10/electric-cars-big-battery-waste-problem-lithium-recycling.

25. Jack Ewing, "What Needs to Happen Before Electric Cars Take Over the World," *New York Times*, December 18, 2017. www.nytimes.com/2017/12/18/business/electric-car-adoption.html.

26. Patrick E. George and Kristen Hall-Geisler, "5 Reasons Not to Buy a Hybrid," How Stuff Works, August 31, 2009. auto.howstuffworks.com/5-reasons-to-not-buy-hybrid5.htm.
27. Josh Goldman, "Electric Vehicles, Batteries, Cobalt, and Rare Earth Metals," Union of Concerned Scientists, October 25, 2017. blog.ucsusa.org/josh-goldman/electric-vehicles-batteries-cobalt-and-rare-earth-metals.
28. Quoted in Michael Parfit, "Future Power: Where Will the World Get Its Next Energy Fix?," *National Geographic*, August 2005. www.nationalgeographic.com/environment/global-warming/powering-the-future/.
29. Chris Smith, "The Trouble with Biofuels," *The Naked Scientists*, April 24, 2013. www.thenakedscientists.com/articles/science-news/trouble-biofuels.
30. Smith, "The Trouble with Biofuels."
31. Jude Clemente, "Why Biofuels Can't Replace Oil," *Forbes*, June 17, 2015. www.forbes.com/sites/judeclemente/2015/06/17/why-biofuels-cant-replace-oil/#3cf1a85ff60f.
32. Quoted in John W. Schoen, "Rising Oil Prices Spark Interest in Biofuels," MSNBC, August 25, 2005. www.msnbc.msn.com/id/7549534/.
33. Freyr Sverrisson, "Missing in Action: Iceland's Hydrogen Economy," Worldwatch Institute, November/December 2006. www.worldwatch.org/node/4664.
34. Lowana Veal, "Hydrogen, Iceland and the Future of Transportation," *InDepth News*, April 14, 2017. www.indepthnews.net/index.php/sustainability/affordable-clean-energy/1074-hydrogen-iceland-and-the-future-of-transport.
35. Quoted in Christine Woodside, *The Homeowner's Guide to Energy Independence*. Guilford, CT: Lyons, 2006, p. 64.

Chapter 3: The Power of the Sun and Wind

36. "How Much Energy Does the Sun Produce?," *Old Farmer's Almanac*, accessed on March 1, 2018. www.almanac.com/content/how-much-energy-does-sun-produce.
37. "Products & Services: How Does It Work?," VSolar Group Berhad, accessed on May 1, 2018. www.vsolar.com.my.

38. "What Is a Watt, Anyway? Understanding Energy and Power Metrics," BuildingGreen, April 20, 2011. www.buildinggreen.com/news-article/what-watt-anyway-understanding-energy-and-power-metrics.
39. "What Size Wind Turbine Do I Need?," *Windpower Engineering & Development*, January 13, 2010. www.windpowerengineering.com/construction/size-wind-turbine-need/.
40. Katie Fehrenbacher, "Solar Is Going to Get Ridiculously Cheap," *Forbes*, June 13, 2016. fortune.com/2016/06/13/solar-to-get-crazy-cheap/.
41. Jackie Craven, "The Importance of Infrastructure," *ThoughtCo.*, December 27, 2017. www.thoughtco.com/what-is-infrastructure-why-important-177733.
42. Ivan Penn, "California Invested Heavily in Solar Power. Now There's So Much that Other States Are Sometimes Paid to Take It," *Los Angeles Times*, June 22, 2017. www.latimes.com/projects/la-fi-electricity-solar/.
43. Amal Ahmed, "We're Doing a Great Job of Creating Renewable Energy—but We Don't Have the Infrastructure to Actually Use It," *Popular Science*, January 18, 2018. www.popsci.com/renewable-energy-infrastructure.
44. Reuters Staff, "German Household Prices at Record High—Verivox," Reuters, October 24, 2017. www.reuters.com/article/germany-electricity-retail/german-household-power-prices-at-record-high-verivox-idUSL8N1MZ30X.
45. Bjorn Lomborg, "How Green Policies Hurt the Poor," *Spectator*, April 5, 2014. www.spectator.co.uk/2014/04/let-them-eat-carbon-credits/.
46. CTBR Staff Writer, "Top Wind Power Producing Countries," *Energy Business Review*, November 3, 2017. wind.energy-business-review.com/news/top-wind-power-producing-countries-031017-5964279.
47. Dominic Dudley, "Renewable Energy Will Be Consistently Cheaper Than Fossil Fuels by 2020, Report Claims," *Forbes*, January 13, 2018. www.forbes.com/sites/dominicdudley/2018/01/13/renewable-energy-cost-effective-fossil-fuels-

2020/#313b998e4ff2.

48. "Barriers to Renewable Energy Technologies," Union of Concerned Scientists, last updated December 20, 2017. www.ucsusa.org/clean-energy/renewable-energy/barriers-to-renewable-energy#.Wp679WXse-I.
49. Quoted in Penn, "California Invested Heavily in Solar Power."
50. Courtney Balestier, "Coal Is Dying—Coal Country Doesn't Have To: Creating the Post-Coal Economy in Appalachia," *Fast Company*, December 12, 2016. www.fastcompany.com/3065766/coal-is-dying-coal-country-doesnt-have-to-creating-the-post-coal-economy-in-appalachia.
51. Kevin Hardy and Donnelle Eller, "6 Common Complaints Against Iowa Wind Turbines," *Des Moines Register*, April 21, 2017. www.desmoinesregister.com/story/money/business/2017/04/21/6-common-complaints-against-iowa-wind-turbines/100706178/.
52. Quoted in Felicity Barringer, "Debate over Wind Power Creates Environmental Rift," *New York Times*, June 6, 2006, p. A18.
53. Julia Layton, "Do Wind Turbines Cause Health Problems?," How Stuff Works, November 5, 2008. science.howstuffworks.com/environmental/green-science/wind-turbines-health1.htm.
54. Keith Kloor, "Can Wind Turbines Make You Sick?," *Slate*, March 20, 2013. www.slate.com/articles/health_and_science/alternative_energy/2013/03/wind_turbine_syndrome_debunking_a_disease_that_may_be_a_nocebo effect.html.

Chapter 4: Using the Earth's Resources

55. Dave Levitan, "Why Wave Power Has Lagged Far Behind as Energy Source," *Yale Environment 360*, April 28, 2014. e360.yale.edu/features/why_wave_power_has_lagged_far_behind_as_energy_source.
56. Levitan, "Why Wave Power Has Lagged."
57. "Hydropower Explained: Tidal Power," U.S. Energy Information Administration, last updated March 22, 2017. www.eia.gov/energyexplained/index.cfm?page=hydropower_tidal.

58. "Geothermal Energy," *National Geographic*, accessed on April 12, 2018. www.nationalgeographic.com/environment/global-warming/geothermal-energy/.
59. "Geothermal Energy," *National Geographic*.
60. Patrick Moore, "Going Nuclear: A Green Makes the Case," *Washington Post*, April 16, 2006. www.washingtonpost.com/wp-dyn/content/article/2006/04/14/AR2006041401209.html.
61. Quoted in Damon Embling, "The EU's Nuclear Power Dilemma," *Euronews*, last updated April 11, 2018. www.euronews.com/2018/04/11/the-eu-s-nuclear-power-dilemma.

Chapter 5: What Happens Next?

62. Heather Kohn, email interview by author, April 16, 2018.
63. Kohn, email interview.
64. Todd Myers, "U.S. Mayors on Climate: This Time Will Be Different," *National Review*, June 15, 2017. www.nationalreview.com/2017/06/us-mayors-climate-pledges/.
65. Klare, *Blood and Oil*, p. 187.
66. David Nield, "This New Hybrid Solar Cell Can Harvest Electricity from Actual Raindrops," Science Alert, March 9, 2018. www.sciencealert.com/hybrid-solar-cell-captures-energy-from-sun-and-rain.
67. Julia Rosen, "Vast Bioenergy Plantations Could Stave Off Climate Change—and Radically Reshape the Planet," *Science*, February 15, 2018. www.sciencemag.org/news/2018/02/vast-bioenergy-plantations-could-stave-climate-change-and-radically-reshape-planet.

DISCUSSION QUESTIONS

Chapter 1: Why Is Alternative Energy Necessary?

1. What is the "energy economy"?
2. How does reliance on fossil fuels affect the economy? How does it affect the environment?
3. What is your opinion of fracking and offshore drilling?

Chapter 2: Rethinking Transportation

1. What are some roadblocks to alternative fuels gaining broad public appeal?
2. Would you want to own a hybrid or electric car? Why or why not?
3. What alternative vehicle fuel do you believe holds the most promise? Why?

Chapter 3: The Power of the Sun and Wind

1. How do solar thermal systems differ from systems using photovoltaic cells?
2. What are some of the factors that must be taken into account when using wind power?
3. What are the economic and political barriers to widespread use of solar and wind power?

Chapter 4: Using the Earth's Resources

1. Describe the benefits and drawbacks of generating power from ocean waves and tides.
2. Do you think nuclear power is a good source of energy? Why or why not?
3. Can you think of alternative energy sources that were not discussed in this chapter?

Chapter 5: What Happens Next?

1. What would be the pros and cons for a consumer when considering switching to green power?
2. Do you think Americans are capable of changing their attitudes about energy?
3. Do you have any ideas for a new or improved energy system?

ORGANIZATIONS TO CONTACT

American Solar Energy Society
2525 Arapahoe Avenue, Suite E4-253
Boulder, CO 80302
(303) 443-3130
info@ases.org
www.ases.org/

This nonprofit organization is devoted to the development of renewable energy. Its website includes a link to the online version of *Solar Today* magazine, with articles on advances in solar technology and other forms of alternative energy.

American Wind Energy Association (AWEA)
1501 M St. NW, Suite 900
Washington, DC 20005
(202) 383-2500
www.awea.org

The AWEA represents the wind power industry and provides information on all aspects of wind power. Its website contains a resource library with detailed fact sheets, answers to frequently asked questions, and ways people can get involved in advocating for wind power.

Energy Information Administration (EIA)
1000 Independence Ave. SW
Washington, DC 20585
(202) 586-8800
infoctr@eia.gov
www.eia.doe.gov

This U.S. government agency offers a wide variety of energy-related information, including information on both renewable and nonrenewable energy sources.

National Resources Defense Council (NRDC)
40 West 20th Street, 11th Floor
New York, NY 10011
(212) 727-2700
nrdcinfo@nrdc.org
www.nrdc.org

The NRDC is an environmental action organization working on a variety of issues, including oil dependence and climate change.

Sierra Club
2101 Webster Street, Suite 1300
Oakland, CA 94612
(415) 977-5500
information@sierraclub.org
www.sierraclub.org

The Sierra Club is an environmental organization with campaigns that encourage replacing fossil fuels with alternative energy. The organization has a network for young adults called the Sierra Student Coalition, which encourages high school and college students to get involved in working toward environmental, climate, and social justice.

FOR MORE INFORMATION

Books

Gore, Al. *An Inconvenient Truth: The Crisis of Global Warming*. New York, NY: Viking, 2007.

Environmentalist and former vice president Gore's best-selling book about global warming has been adapted for a young adult audience. Gore's book details which human activities have aided in the destruction of the environment and what will happen if no actions are taken.

Green, Robert. *How Renewable Energy Is Changing Society*. San Diego, CA: ReferencePoint Press, 2016.

As different forms of energy replace fossil fuels, some aspects of society must change too. Green discusses how these two areas are connected.

McPherson, Stephanie Sammartino. *Arctic Thaw: Climate Change and the Global Race for Energy Resources*. Minneapolis, MN: Twenty-First Century Books, 2015.

Due to melting ice and permafrost as a result of global warming, fossil fuels and other natural resources that were previously inaccessible can now be reached. Learn about how the battle for these resources and the process of extracting them from the earth are affecting ecosystems in the Arctic.

Sawyer, Ava. *Human Environmental Impact: How We Affect Earth*. North Mankato, MN: Capstone Press, 2018.

The author explores the impact human life has on various aspects of the environment.

Small, Cathleen. *Wind, Waves, and the Sun: The Rise of Alternative Energy*. New York, NY: Cavendish Square Publishing, 2018.

This book discusses the history of alternative energy as well as the state of the field today.

Websites

Environmental Literacy Council
www.enviroliteracy.org
The Environmental Literacy Council is dedicated to helping students understand environmental issues. Its comprehensive website contains resources on almost every aspect of the natural world, including detailed information on energy issues.

***National Geographic*: You Have the Power**
www.nationalgeographic.org/game/national-geographic-energy/
This interactive game allows people to see the numerous decisions that officials have to consider while making decisions about energy types. Players can choose which combination of energy sources should be used to provide energy to towns in different environments. Try to provide 100 percent energy to the town without going over budget and while keeping the concerns of the towns' citizens in mind.

National Renewable Energy Laboratory (NREL)
www.nrel.gov
NREL is committed to researching creative solutions to the energy crisis.

350
350.org
This nonprofit organization is dedicated to replacing fossil fuels with 100 percent renewable resources.

Union of Concerned Scientists
www.ucsusa.org
This website contains articles in English and Spanish about different types of alternative energy sources as well as the environmental issues surrounding energy use.

U.S. Department of Energy
energy.gov
This government website provides detailed information on clean energy sources, scientific energy innovations, climate change, and price trends in the energy economy.

INDEX

PICTURE CREDITS

Cover hfzimages/Shutterstock.com; p. 7 Nickolay Khoroshkov/Shutterstock.com; p. 8 Vyacheslav Svetlichnyy/Shutterstock.com; p. 12 jgorzynik/Shutterstock.com; p. 13 huyangshu/Shutterstock.com; p. 16 AP Photo/David Goldman; p. 17 Hung Chung Chih/Shutterstock.com; p. 20 FloridaStock/Shutterstock.com; p. 23 seeshooteatrepeat/Shutterstock.com; p. 26 EQRoy/Shutterstock.com; p. 29 John B. Carnett/Bonnier Corp. via Getty Images; p. 30 TigerStock's/Shutterstock.com; p. 32 GRIFF TAPPER/AFP/Getty Images; p. 35 Mohammed Talatene/Anadolu Agency/Getty Images; p. 38 igorstevanovic/Shutterstock.com; p. 41 Nicole S Glass/Shutterstock.com; p. 45 Worldpics/Shutterstock.com; p. 46 Waldo Swiegers/Bloomberg via Getty Images; p. 50 smartflower energy technology GmbH/obs via AP Images; p. 52 XXLPhoto/Shutterstock.com; p. 60 Citizens of the Planet/Education Images/UIG via Getty Images; p. 61 Ted Soqui/Corbis via Getty Images; p. 65 Holger Leue/Lonely Planet Images/Getty Images; p. 68 Holly Kuchera/Shutterstock.com; p. 71 Jemastock/Shutterstock.com; p. 72 Breedfoto/Shutterstock.com; p. 74 dvoevnore/Shutterstock.com; p. 77 Inspired By Maps/Shutterstock.com; p. 78 DedMityay/Shutterstock.com; p. 83 William Perugini/Shutterstock.com; p. 84 Chones/Shutterstock.com.

ABOUT THE AUTHOR

Sophie Washburne has been a freelance writer and editor of young adult and adult books for more than 10 years. She travels extensively with her husband, Alan. When they are not traveling, they live in Wales with their cat, Zoe. Sophie enjoys doing crafts and cooking when she has spare time.